D1044945

How to HOUSEPLANT

How to
HOUSEPLANT

A Beginner's Guide to Making and Keeping Plant Friends

HEATHER RODINO

STERLING
New York

STERLING
New York

An Imprint of Sterling Publishing Co., Inc.
1166 Avenue of the Americas
New York, NY 10036

STERLING and the distinctive Sterling logo are registered trademarks
of Sterling Publishing Co., Inc.

Text © 2019 Heather Rodino

All rights reserved. No part of this publication may be reproduced,
stored in a retrieval system, or transmitted in any form or by any means
(including electronic, mechanical, photocopying, recording, or otherwise)
without prior written permission from the publisher.

ISBN 978-1-4549-3290-1

Distributed in Canada by Sterling Publishing Co., Inc.
c/o Canadian Manda Group, 664 Annette Street
Toronto, Ontario M6S 2C8, Canada
Distributed in the United Kingdom by GMC Distribution Services
Castle Place, 166 High Street, Lewes, East Sussex BN7 1XU, England
Distributed in Australia by NewSouth Books
University of New South Wales, Sydney, NSW 2052, Australia

For information about custom editions, special sales, and premium and
corporate purchases, please contact Sterling Special Sales at 800-805-5489
or specialsales@sterlingpublishing.com.

Manufactured in China

8 10 9

sterlingpublishing.com

Interior design by Shannon Nicole Plunkett
Cover and endpaper design by David Ter-Avanesyan

For image credits, see page 173

To my mom, Rowena Rodino,
who cleans the scale off my plants
when she comes to visit

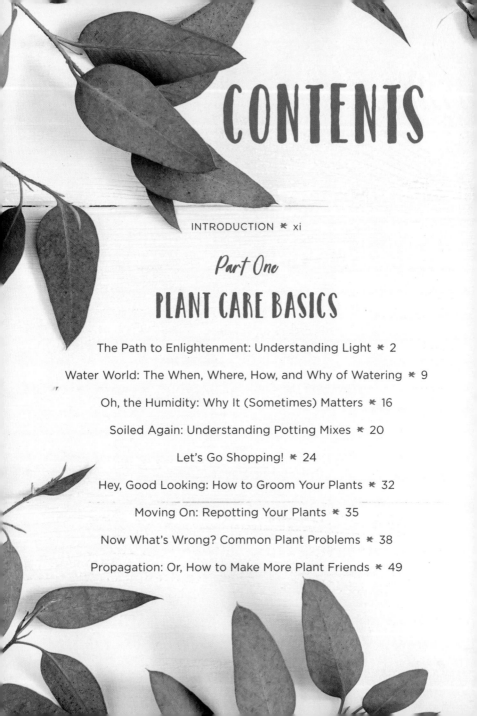

CONTENTS

Part Two

PLANT PROFILES

INTRODUCTION

I t's easy to fall in love with houseplants and the quiet joy their presence can bring. They're beautiful companions, certainly, but they also quite literally add more life to your space, making your home feel more like your own, whether you're in a small rented apartment where you can't paint the walls, or in a sprawling house. Houseplants can provide a form of reprieve and respite from the stress of everyday life and can improve your indoor air quality, your mood, and even your health. Fortunately, there's no one-size-fits-all approach to keeping houseplants. You can dote on them as you would with a favorite pet; focus more on their aesthetics, using them to enhance any kind of interior design; or look at them as a way of bringing the outdoors inside.

What's more, houseplants don't need to be a challenge. Really, anyone can be a good friend to a houseplant, if provided with the right information. The first thing you have to put out of your head, even if you've had bad luck before, is that there's such a thing as a black thumb. It's normal, however, that some questions should arise. How do I choose a plant? How much light does my plant need? How often do I have to water it? What happens if I forget? Is fertilizer necessary, and how often should I use it? How can I tell when I need to repot my plant? What kind of soil should I use? Why are the leaves turning brown, and where did these bugs come from? If you've ever wondered about the answer to any of these questions, you've come to the right place. Even if you don't know what questions to ask and feel like you're doomed to kill every plant that crosses your doorstep, this book will take you step by step through the basics, teaching you how to care for—and love!—your houseplants.

Like many people, I struggled with keeping plants alive and thriving for a long time. I'd go to the store, find a plant (okay, more like five plants) that I liked, and come home with these new friends. The plants would look great—for a while. Without fail, a few months later I'd have some casualties. What was I doing wrong? I decided to look into the matter more seriously and earnestly than I had before, to teach myself about plants.

The most important thing I discovered along the way seems obvious, but it might not be to everyone (it wasn't to me!): learn about each plant you're interested in and how it might work in your home. In the following pages, you'll learn how to assess your home and the environment it offers, so you can select your plants accordingly. And if you've got your heart set on a plant that might not work in your space, you'll go in knowing that, so you can limit your disappointment if it fails to thrive. (But who knows, a little passion can go a long way toward success.)

You'll also learn the benefits of spending a little time with your plants on a regular basis, which will help you understand their needs

The basics of plant care are easy to learn.

a bit more. This simple, mindful practice will teach you more about plants than you could ever imagine—you'll become an expert on how to take care of *your plants* in the unique microclimate called *your home.* You'll discover, for example, that not all of your plants will need water at the same time. You'll find diseases and pests early on and be able to treat them more successfully. You'll breathe on your plants too, exhaling the carbon dioxide that they need. As a bonus, you may find that stopping by to visit your plants can be a way of checking in with yourself and stepping away from technological distractions. When (not if) you succeed, and a plant starts to get new leaves or begins to flower, you'll get an incomparable rush of accomplishment that will have you beaming with pride and wanting more.

If you've picked up this book, you're already one step closer to that moment. In the following pages, you'll learn the basics of plant care and find profiles for fifty well-loved, time-tested houseplants, including some new favorites. My hope is that the foundational information offered here will give you the tools and confidence you need to cultivate your own plant collection—to help you learn how to houseplant!

Part One
PLANT CARE BASICS

THE PATH TO ENLIGHTENMENT

Understanding Light

When it comes to plants, most of us have never given the light in our home a second thought, but the type of light that your home receives is perhaps the most important factor to consider when choosing your houseplants. After all, you can adjust your watering habits or humidity levels, but it's hard to build a south-facing window if you don't already have one. Give yourself some time to understand the light in your home and how your plants respond to it. You may have your heart set on a fiddle-leaf fig (page 96), but if you can't provide it with bright light, you've set yourself up for an expensive disappointment.

Before you select plants for your home, let's look at why they need light. Light is "food" for plants. It allows them to undergo *photosynthesis*, a process during which the plant *synthesizes* its own food using light (*photo* comes from the Greek root *phōs*, meaning "light"), water, and carbon dioxide in the air. These components are converted to glucose (sugar), and the plant releases the oxygen from the water molecule into the air.

Admittedly for indoor gardening, light can be one of the most confounding components of plant care. The light requirements on the plant's tag, if it even has one, are often tersely worded, so how do you really know if you can provide the kind of light needed to grow a particular plant successfully?

The amount of light your home receives depends on many factors, including the direction your window is facing, the size of the window

An aloe plant (page 64), Chinese money plant (page 82), and *Oxalis triangularis* (page 128) sit happily on a windowsill that receives indirect light.

(and how clean it is!), where you live, the time of year, the hours of daylight on a given day, and the obstacles outside your window, such as other buildings or trees. The best way to understand where to put your plants is to learn a bit about the light available from each exposure (north, south, east, and west), pay attention to the conditions in your home, and observe your plants' behavior after you put them in place. Don't be afraid to experiment here; it can take a bit of practice! A plant that leans strongly toward the light—a process called *phototropism*—may not be getting enough. Conversely, like humans, plants can get too much sun and become sunburned. Some plants, however, can be gradually acclimated to higher light situations (see More Sun, Please!, page 8).

Let's look at the light available from the north, south, east, and west. If you don't know which way a window faces, take out your smartphone, open the compass app, and find out. Or simply pay attention to where the sun rises!

Many houseplants, such as this jade plant (page 108), will lean toward the light source. To promote even growth, give them a quarter-turn when you water them. Phototropism can also be an indication that your plant needs more light.

PLANTS AND HEALTH

Thanks to oft-cited studies by NASA and other organizations, we now know that many plants can improve air quality, reducing the volatile organic compounds (VOCs), such as formaldehyde, benzene, and trichloroethylene, in indoor air. (Linked to a range of health problems, VOCs can be found in carpet, upholstery, paints, cleaning products, aerosol sprays, and other commonly used items in the home.)

Houseplants can also improve well-being. One small 2015 study found that interacting with indoor plants can measurably reduce psychological and physiological stress, and a 2009 study found that plants enhanced the outcomes of surgery patients. In prisons, retirement homes, juvenile detention centers, and veterans' homes, horticultural therapy is growing as a practice to help those dealing with post-traumatic stress disorder, anxiety, depression, and other issues.

NORTHERN EXPOSURE

A window facing the north, or northern exposure, receives only indirect light. Northern exposure is generally considered low light. (An exception would be if you have a bay window, which would also provide some eastern and western exposure.) Most plants do poorly with only northern exposure, particularly in winter, but there are still a few great ones for you to choose from. While you might not succeed with a Meyer lemon tree (page 116), pothos (page 140), philodendrons (pages 103 and 154), and the Victorian-era darling, the cast iron plant (page 78), among others, should all work well for you. You can even grow a breathtaking lady slipper orchid (page 110). The key is to get them as close to the light source as possible. White walls and mirrors that reflect the available light can help improve the situation. If you've really got the houseplant bug, you can also supplement northern light with fluorescent light or grow lights to increase the range of plants you can grow.

The Upside Down

For our friends in the Southern Hemisphere,
northern exposure provides direct sun,
while southern exposure provides no direct light.

SOUTHERN EXPOSURE

You've probably heard of southern exposure. It's much lauded in real estate, but what does it mean for plants? In general, if you have south-facing windows, you can grow plants that need full sun. Plants receiving southern exposure get bright, strong light all day long. Most cacti, succulents, and citrus will thrive, but for many plants this is too much light. In that case, you can place plants back a few feet from the window or put up a sheer curtain or blinds, which will give you bright filtered light. During shorter winter days when the light is less intense, you can experiment with giving your plants southern exposure, regardless of their light requirement. You may find that they do better.

Blinds and sheer curtains can make even the sunniest window more hospitable for plants that might not thrive with long periods of direct sunlight.

EASTERN EXPOSURE

Plants in east-facing windows will receive nice bright morning light as the sun rises; the exact number of hours will vary, depending on the time of year. In the afternoon, your plants will get a break and receive indirect light. Many plants that need bright light but not full sun will be happy with eastern exposure, and those with low to moderate light requirements will do well. Eastern exposure is also ideal for plants that prefer cooler temperatures and are prone to sunburn, as it is not as warm as western exposure.

WESTERN EXPOSURE

What does the west have to offer? Your plants will not get direct morning light, but the afternoon sun will be hotter and more intense than the morning sun. Plants with higher light needs will do well with western exposure, especially when placed on a windowsill. Many other plants that have bright or indirect light requirements will do well if they are set back a bit from the window. Western exposure is also beneficial for plants that like to be a little toastier.

MORE SUN, PLEASE!

As you read through the growing recommendations in the plant profiles, you may notice that some indicate that you can acclimate plants to brighter exposures, such as southern or western, or even outdoors. To do so, place your plant in the new location for a few hours every day, gradually increasing the time until it's comfortable in that spot full time. This process may take several weeks. If the leaves start to get brown spots (an indication of sunburn) or start to wilt, acclimate the plant more slowly.

WATER WORLD

The When, Where, How, and Why of Watering

N ow that you know where to place your plants, how do you take care of them once they're there? Let's look a little closer at the next crucial aspect of care: watering. Each plant has its own watering requirements. When you buy new plants, get to know what they are and don't be afraid to adapt them to the conditions in your home.

WHEN TO WATER

It would be great if you could just water your plants once a week and be done with it. But with this strategy, you might not be watering enough for some plants, like ferns, and overwatering other plants, like cacti and succulents. Temperature, humidity, and the seasons can affect water requirements. For example, plants may need less frequent waterings in a cool, humid room than in a warm, dry room. You may find yourself watering more often in summer when the room temperature is higher than in winter when some plants go dormant. Conversely, other plants may dry out more quickly in winter when the heat is running. Even the kind of pot can have an effect. Plants in plastic pots retain water longer than plants in porous terra-cotta pots. With all these factors coming into play, what's the secret?

The trick to successful watering is to *check* whether your plants need water, not just water them automatically. Most plants do well if you let the top of the soil dry out slightly before watering, usually to a depth of about an inch (2.5 cm), but check the plants' individual requirements. Another way to see if your plants need water is to pick up the pot and gauge the weight. It will feel much lighter when it needs water as compared to right after a thorough watering. It takes a bit of practice to get a sense of how much the pot usually weighs.

It's a good idea, especially when you are first starting out, to check on your plants' water needs most days. If you tend to forget, consider setting up a smartphone alert to remind you. Instead of sitting down immediately in front of your computer or tablet in the morning, grab your mug of coffee or tea and visit your botanical buddies. Or check on them at the end of the day, as a way of unwinding, of disconnecting from the human world and reconnecting with the natural one.

HOW TO WATER

When you water, be generous. Pour water into the potting mix until it comes out of the drainage holes in the bottom of the container and collects in the saucer, making sure to water around the whole plant, not

Overwatering is the number-one killer of plants.

Almost no plant wants to sit in soggy soil; it can lead to a fungal infection called root rot (page 44) that will eventually kill your plant. Inconsistent watering can also stress out a plant, making it more vulnerable to pests and other diseases.

---

just in one spot. (This technique also applies to plants that don't require a lot of water in terms of frequency, like cacti and succulents.) After the excess water has had a chance to collect in the saucer, dump it out. You're done! If you are using a pebble tray to boost humidity levels around the plant (see Oh, the Humidity, page 16), it's fine for the water to remain in the saucer as long as the base of the pot is not sitting in the water.

WATER QUALITY AND TEMPERATURE

Tap water is fine for most plants, but a few plants are sensitive to hard water, chlorine, and fluoride. For such plants, some experts suggest giving such plants rainwater, distilled water, or filtered water. If you have a way of collecting rainwater, that's great, but filtered water will keep them happy and is much easier to manage. Temperature matters, too: stick with tepid water.

SPECIAL WATERING PRACTICES

Most plants are watered from the top, meaning that water is poured directly into the soil. (This is called *top watering*.) Certain plants, however, benefit from special watering practices. African violets (page 60) are commonly watered from the bottom (*bottom watering*)

A watering can with a long, narrow spout makes it easy
to direct the water exactly where you want it.

because they have sensitive leaves that can discolor and rot if water sits on them. To water from the bottom, place the plant pot in a dish with an inch or two (2.5–5 cm) of water. Leave it there for about half an hour to allow the plant to absorb water through the drainage holes. You can soak moth orchids (page 121), which usually grow in a potting medium made of chunks of bark, in a basin of room-temperature water for 20 to 30 minutes about once a week. Other plants, like bromeliads, have a central "vase" that you pour water into. Still others, like a bird's nest fern (page 71), would rot if you poured water into the center of the plant and prefer watering directly in the soil. Again, check each plant's requirements for specifics.

Plants benefit from good air circulation, which can help prevent fungal problems.

Open your window on a warm day (no cold drafts, please!), turn on a ceiling fan, or even run a small desk fan. Avoid placing your plants in a spot where they could receive a blast of air from a heating vent or air-conditioning unit.

FERTILIZING

Many potting mixes contain fertilizers that help supply the nutrients plants need, the ones that they would receive naturally from decaying matter if they were growing outdoors in soil. Over time, however, your plants will deplete the nutrients in the potting mix, which is why fertilizer is a key component for keeping your plant friends happy and healthy.

Just as you don't want to be overzealous with watering your plants, you don't want to fertilize them excessively. More is definitely not more. Overfertilizing can cause brown leaf tips, yellowing leaves, slow growth, and other problems. However, when properly—and conservatively—applied, fertilizer can correct nutrient deficiencies and help your plants thrive.

An all-purpose organic liquid houseplant fertilizer is an excellent choice. You simply add a dose directly to the water in your watering can. It will work for most of your plants, though you can get specialized fertilizers for citrus trees, orchids, African violets, and other kinds of plants. If you choose a nonorganic fertilizer, which will typically come in a water-soluble formula, pay attention to the number on the label. It indicates the balance of nutrients in the fertilizer—nitrogen, phosphorus, and potassium, respectively. Select one that is balanced, meaning that it contains equal parts of all three nutrients (e.g., 5-5-5 or 10-10-10). For a flowering plant, however, you can use a fertilizer with a higher phosphorus number (e.g., 15-30-15) for bigger, longer-lasting blooms.

Fertilizer comes in other forms as well. Granules can be spread on top of potting mix. Slow-release spikes can be inserted in the potting mix, but take note: they can fertilize unevenly.

Here are a few guidelines for using fertilizer:

- Use no more than half of what the label suggests. This will help reduce the buildup of fertilizer salts, which can burn the plants' roots.

- In general, fertilize plants about once a month during spring and summer when they are actively growing. Reduce fertilizer in fall, and in most cases skip the fertilizer entirely in winter.

- A few times a year, take your plants to the sink and water them generously from there to flush out any accumulated fertilizer salts.

- If your potting mix contains fertilizer, wait at least a month after repotting before giving the plant more fertilizer. Likewise, when you buy a new plant, hold off on the fertilizer for a few weeks while it adjusts to its new home. It may have been fertilized recently.

- At first it may seem counterintuitive, but don't feed a plant that has pests or diseases. It may do more harm than good. Instead, correct the underlying problem first, and then fertilize once the plant recovers.

OH, THE HUMIDITY

Why It (Sometimes) Matters

Humidity is moisture in the air. Some plants love it. Others don't care one way or the other. It's important to know which plants need it, so you can make your home as welcoming as possible. If the air is too dry, the leaf tips of your plant may turn brown and flower buds may drop off before blooming. The care requirements for all plants in this book indicate whether supplemental humidity is helpful.

I had always found humidity requirements a bit of a stumbling block when it came to plant success, but it's easy to understand and not that hard to create if you don't have it.

Here's a rule of thumb (there are, of course, exceptions): Plants with thin, delicate leaves (such as ferns, nerve plant, and polka dot plant) and tropical plants like humidity. Most cacti and succulents, which have plump leaves, stems, or stalks, don't need it. Think of tropical conditions: warm and steamy. Compare this with desert conditions, which are usually hot, dry, and sunny. (Jungle cacti do, however, like humidity.)

While many indoor plants would be perfectly happy with 70 to 80 percent humidity in the home, such levels are not typically realistic—or comfortable for the home's human inhabitants. Fortunately, levels of 50 to 60-plus percent will also keep your plants comfortable, and most will tolerate less-than-optimal conditions. I have a digital thermometer with a hygrometer that gives me both the indoor temperature and the humidity percentage, but there are other ways of determining humidity levels. Your climate is one clue. Do you live in a desert region where the air is reliably dry? Or perhaps you live in a subtropical area where the humidity is on the higher side. Heating and air conditioning will dry out the air in your home. Take a look at your skin—does it seem dry as well? If you know your home generally has dry air and you don't want to worry about whether your plant is getting enough moisture, simply choose plants that do well in dry air. You'll find this information listed in the plant profiles.

ROOM (TEMPERATURE) TO GROW

Most houseplants come from tropical or subtropical locations, which makes them a great fit for the temperature-controlled environment we can offer in our homes. In general, average room temperatures, roughly between 60 and 80°F (16 and 27°C), will suit most plants, and they'll even be fine at slightly cooler or warmer temperatures. For more specific suggestions, visit the individual plant profiles.

THE BIG DEBATE: TO MIST OR NOT TO MIST?

Some gardening books and experts will tell you that you must mist your plants to give them extra humidity. Others will say that there's no point in misting because the effects are only momentary, and it might even lead to diseases in plants with sensitive leaves. What's a confused plant novice to do? I personally don't mist my plants, but then I live in a humid environment. If you live in a dry place and are growing plants with high humidity requirements, go ahead and mist. Misting can also be a good idea in winter, when the air is dry. But before you go spritzing everything in sight with your vintage brass mister, check the plant's individual humidity needs. For example, you'll want to go gently with African violets (page 60)—or even skip misting them entirely—because water left on the leaves can cause stains; but you can mist away at your humidity-loving Boston fern (page 74).

Here are several easy ways to provide more humidity for your plants.

Watch your watering schedule. If the air is particularly dry, your plants might need water more frequently than if they were in a more humid environment.

Use a pebble tray. It's very simple to make a pebble tray. Fill a saucer with some pebbles and add water, making sure it doesn't rise above the level of the pebbles. The plant should be sitting on top of the pebbles, not in the water. As the water evaporates, it will increase humidity levels around the plant. Periodically clean the saucer to prevent mold and mildew buildup.

Pebble trays are easy to create on your own. Just place a handful of pebbles and small rocks in a tray or saucer.

Make a plant gang. By placing several plants together, you will automatically increase the humidity level as the plants form their own microclimate.

Run a humidifier. If you have a number of plants that prefer higher humidity levels, consider running a humidifier in the room. You may find that it's beneficial to you as well!

Group a nerve plant with some moss and ivy for an eye-catching terrarium that can fit on a shelf or a corner of a desk.

Make a terrarium. Some humidity-loving plants thrive when grown in terrariums. Plus they look super cute! You'll find plenty of inspiration online.

Move plants to a more humid room. Assuming that you have sufficient light in your bathroom or kitchen, place plants with higher humidity needs there.

SOILED AGAIN

Understanding Potting Mixes

What is potting mix? For starters, it doesn't have any soil in it. Instead, it is a blend of several ingredients (see In the Mix on page 22 for more), the exact combination of which varies by manufacturer. It is designed to provide the proper balance of water retention, aeration, and drainage, while allowing the plants' root systems to grow. If you watch gardening programs or read about gardening, you might have also heard it referred to as a "soil-free mix" or "potting medium."

It's important to note that you can't just go outside, dig up some dirt, and dump it into your pot. It's fine for outdoor gardening, where worms and other creepy-crawlies can aerate the soil, but in a container, outdoor soil becomes impossibly dense and hard—not exactly what you're looking for in a growing medium. It can also contain pests and other unhelpful organisms. In addition, don't make the mistake of picking up something labeled "garden soil."

It's meant for outdoor use, not containers. And whenever possible, choose organic.

When you're just getting started, I recommend sticking with a pre-mixed potting medium. While you can certainly mix up your own, you might start to have second thoughts about the whole houseplant thing if you have to trudge over to the garden center, buy several huge bags of mysterious substances, and then be expected to mix two parts of this, one part of that, and one part of the other thing on the floor of your apartment. Instead, let's look at some of the kinds of premade potting mix you might need.

ALL-PURPOSE POTTING MIX

The name says it all. This is the mix you'll use the most. It may be labeled "all-purpose potting mix" or simply "potting mix." For most plants, choose this option. If the plant needs a little extra drainage (check the soil requirements), you can always stir in some extra perlite, vermiculite, or sand.

Many potting mix formulas include perlite, which appears as small white granules.

IN THE MIX:
Understanding Your Potting Medium

What's in the special sauce? Your potting mix will contain some combination of the following.

BARK (OR BARK FINES) Very small particles of composted crushed or shredded bark, often from pine.

COIR A shredded fiber made from coconut husks that holds moisture well and is an environmentally friendly and sustainable alternative to peat moss.

LIMESTONE A rock that is ground up and added to adjust the pH of a peat moss–based potting mix.

PEAT MOSS (ALSO CALLED SPHAGNUM PEAT MOSS) AND/OR HUMUS Decomposed organic matter from the lower levels of peat bogs. It helps retain nutrients and moisture. Peat humus is more fully decomposed than peat moss. (Many gardeners have switched to coir due to environmental and sustainability concerns about peat.)

PERLITE Lightweight little white balls that look like foam but are actually heated volcanic glass. Perlite improves aeration and drainage.

SAND Added to potting mixes to improve aeration and drainage.

SPHAGNUM MOSS The living moss that has been harvested from the surface of a bog and dried. Over a period of thousands of years, it will decompose to make peat moss. Often used as an orchid potting medium.

VERMICULITE Made when micaceous minerals are heated to a super-high temperature and expand. It helps retain moisture and also drains well, but is used less often than perlite.

WORM CASINGS A rich natural fertilizer made from worm poop. (Don't worry; it doesn't smell bad!)

CACTUS AND SUCCULENT MIX

Cacti and succulents don't like their roots to sit in water, so they need a quick-draining mix. A cactus and succulent mix often has added river sand to promote drainage. You can also use this mix for palms. After all, palms often grow on the beach in nature!

ORCHID AND BROMELIAD MIX

Bromeliads and many orchids are *epiphytes*, which are essentially air plants. Epiphytes do not grow in soil, but instead gather their nutrients from the air and rainwater. They often grow on other plants, such as in the crotch of a tree. To grow epiphytes successfully in a pot, they need the very quick-draining and aerating medium that an orchid or bromeliad mix provides. This mix is usually made from chunks of bark, big pieces of perlite, and perhaps charcoal pieces. You can also use coco chips, which are made from coconut husks and typically require rehydration. If you're someone who doesn't like to get your hands dirty, you'll love orchid mix!

Orchid bark, a specialized potting medium consisting primarily of tree bark, comes in three main sizes: fine, medium, and coarse (pictured here).

LET'S GO SHOPPING!

You've evaluated the light and available humidity in your home and understood the kinds of potting mix you might need. Before you head to a nursery to select your plants, I have just a few more tips and things for you to think about.

If you have limited space for plants, stick with a few small ones to brighten up a windowsill, table, or shelf. Don't forget the ceiling! You can choose plants in hanging baskets or pots as a way to maximize your space. If you have a big, open room, consider larger floor plants and trees, which can completely transform a room. Do you want a lush, tropical feel, or would you prefer a plant that stays fairly tidy? Do you like the idea of a plant with trailing stems, like pothos (page 140) or English ivy (page 94)? Or maybe you prefer flowering plants as a way of injecting color into your space.

You should also evaluate your lifestyle and choose plants that will complement it, not cramp it. If you're a nurturer, choose plants that like attention. If you want plants you can mostly forget about except for occasional watering, succulents and cacti are excellent choices. The Greatest Houseplant Hits section (pages 54–57) is the perfect place to start your search, and the plant profiles in this book include options to suit all of these needs.

Ready? It's time to go plant shopping!

Many garden centers and nurseries specialize in houseplants and offer a wider selection than big-box stores.

WHERE SHOULD I BUY MY PLANTS?

Ideally, you want to go to a garden center or nursery, where you can get help from well-informed, educated plant experts who know what they're selling and can answer any question you have. At a big-box store, the employees might not know much about the plants for sale, but you may find plenty of healthy plants from such places.

If you're looking for a specific plant, and your local nursery doesn't carry it, check online. There are numerous established, reputable online sellers that will ship live plants to you. (Check the resources section on pages 168–169 for a few options.) Look for one with a money-back guarantee if you're not satisfied or if something happens to the plant in transit.

HOW DO I CHOOSE A PLANT?

When you pick out a plant, make sure you choose one that's healthy. The plant should look perky, not droopy. Avoid plants with yellowing or browned leaves or stems. If it's a succulent, the leaves should feel firm, not deflated and wrinkly. Look under leaves to make sure there is no sign of bugs or disease. Check the soil for any sign of infestation as well. Not only may a sick plant croak, wasting the money you just spent, but some plant infestations and diseases can spread, infecting your healthy plants at home. Read through Now What's Wrong? on pages 38–48 to learn how to identify common plant problems and ensure that a plant you purchase doesn't have one of these issues.

If a certain plant catches your eye and you're not sure what kind of care it needs, don't be afraid to ask for help. If you can't provide the environment the plant needs to thrive, it's probably best to leave it behind for someone else—or manage your expectations in case it doesn't work for you.

After you've made your purchase, plant care begins even before you leave the shop. Many plants can be quite sensitive to cold conditions, so if you're shopping in winter, make sure your new friend is properly

WHAT'S IN A (BOTANICAL) NAME?

When we talk about plants, we usually call them by their common names: "oak trees," "African violets," "moth orchids," and so on. However, such names have limitations. For example, common names can vary regionally: one person's "arugula" is another's "rocket." What's more, a single common name can refer to multiple genera and species. Take the "money plant." But which one? *Lunaria annua, Crassula ovata, Epipremnum aureum, Pachira aquatica*, or *Pilea peperomioides*? All five are sometimes called money plants, but *they are entirely different plants*. It can get confusing.

That's where *botanical names* come in. A botanical name is a two-part Latin or Latinized name—more formally called *binomial nomenclature*—composed of the plant's genus and species. A botanical name is the same wherever you go, whether it's Boston, Brisbane, or Buenos Aires.

Let's look at one example: *Sansevieria trifasciata* 'Laurentii', a kind of snake plant. "Sansevieria" is the genus, which is composed of dozens of species, including "trifasciata." If a *cultivar* (cultivated variety) exists, it will appear in single quotes after the genus and species, for example, 'Laurentii'. If a plant has been crossed with another plant, thus creating a *hybrid*, then you'll see a multiplication symbol, or ×, in the botanical name, as in the case of the Meyer lemon tree: *Citrus × meyeri*.

Do you really need to know the botanical names in order to take care of your plants? Not necessarily, but knowing them can make it easier to research information about plants and to look for a specific one. If you start getting more into plants, you'll probably pick up some botanical names without even realizing it. You might take particular geeky pleasure in knowing your *Philodendron hederaceum* from your *Philodendron bipinnatifidum*.

bundled up (a Kraft paper wrap will work), even for a brief trip outdoors, and get it home as quickly as possible. Some plants, particularly tropical plants, are best purchased during the growing season, i.e., spring and summer. And it probably goes without saying, but I'll say it anyway: you shouldn't leave a plant in a hot or cold car unless you want to kill it.

WHAT OTHER ITEMS MIGHT I NEED?

There are all kinds of tools and gadgets that you can buy, but at the very least you'll want to get the following.

Pots and Saucers

When you purchase your plant, it will probably come in a plastic pot, but you'll likely want to put it in a prettier container to highlight its beauty. Both a plant *and* its container become part of your decor, and a plant can have a completely different feel depending on the kind of pot

Mix and match terra-cotta pots with containers made from other materials to give your plant collection a fun, whimsical look.

it's in. Fortunately, there are many options to choose from. The most common material for a plant container is terra-cotta. You can also find pots made of ceramic, concrete, earthenware, and metal. Don't forget to pick up a matching or coordinating saucer.

The most important feature that a pot needs to have is a hole or holes in the bottom for drainage. If the pot of your dreams does not have a hole in the bottom, you can treat it as a *cachepot*, a decorative container that holds another pot. When your plant needs water, remove it from the cachepot and water it at the sink. While it's still in the sink, let the water drain completely, then return the plant in its pot to the cachepot until the next watering. You can also use a woven or fabric basket in this fashion to hide a less-than-exciting pot.

If your plant's container looks a little too utilitarian for your taste, dress it up with a basket. The handles also make the plant easy to move to another spot.

Trowel

A small trowel or potting scoop will make any transplanting job easier.

Gardening Gloves

Gloves are useful for repotting plants, whenever you're handling potting mix, and for pruning, as some plants release a sap that can irritate the skin.

Basin

If you have the space to store it, a basin is handy for repotting plants neatly, so you don't get potting mix all over your floor. (Alternatively, you can lay out sheets of newspaper on the floor.) If you're adding other ingredients to a potting mix, you can also use a basin to mix them.

Scissors

A good pair of floral or leaf scissors will be useful for grooming plants and light pruning. If you need to cut back branches or stems, garden shears are indispensable. Your tools should be cleaned and sterilized (rubbing alcohol is an effective, inexpensive disinfectant) every time, to avoid transmitting any pest or disease from plant to plant.

Watering Can

Look for a watering can with a narrow spout so you can be precise and targeted in your watering. Some plants, like African violets (page 60), don't want water on their leaves, and other plants will rot if you get water in the center of the plant too often. A removable sprinkler nozzle is helpful for rinsing the leaves of less sensitive plants.

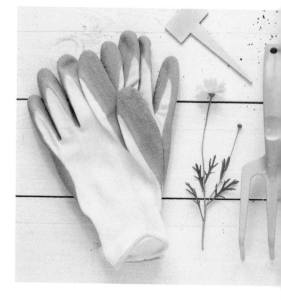

Some gardening gloves, such as the ones shown here, have reinforced sections to protect your fingers from sharp branches. They also make the fingertips of your gloves, which experience the most wear and tear, more durable.

A Word on Toxicity

Many houseplants are toxic to humans and pets
if consumed. Gardening author (and goddess)
Tovah Martin says it best: "None of your houseplants
should be eaten by people or pets." Pets should
be trained to leave the plants alone, and plants
placed out of the reach of small children. If you
are particularly concerned about this issue, start
with the pet-friendly plants listed in Greatest
Houseplant Hits (page 54–57) and visit the
resources section (pages 168–169) for more
information. You'll find links to databases where
you can research individual plants and view
lists of toxic and nontoxic plants.

HEY, GOOD LOOKING

How to Groom Your Plants

J ust like humans, plants need grooming to keep them looking their best. Sometimes a haircut and spa time are in order.

EVERYONE INTO THE TUB!

Indoor plants accumulate dust and dirt on their leaves, which can block their pores and prevent them from photosynthesizing efficiently. If your plants lived outdoors, rain would wash these things away, so to emulate that natural process, I like to give most of my leafy plants a shower to clean the leaves whenever they seem particularly dusty. A shower not only cleans the leaves, but also waters the plant thoroughly and helps flush out excess fertilizer salts. You can take small plants to the kitchen sink and rinse them there. Use the spray function if you have it. If not, a watering can with a sprinkler spout attachment will work. Larger plants can go in the bathroom shower. Only use tepid water for this purpose—never cold, never hot. Once the plants have had a chance to drain well and water is no longer dripping out of their container's drainage holes, return them to their regular location. For plants that are too heavy to move, a soft wet cloth is useful for removing dust and dirt. (Avoid commercial leaf-shine products, which can block leaf pores.)

For plants that don't like to get their leaves wet, you can use a small brush to clean the leaves. Use tweezers or a brush to collect any material stuck in a hairy cactus.

Are the leaves of your plants collecting dust?
Bring them into the tub for a quick shower.

TIME FOR HAIRCUTS!

To make a plant fuller and more symmetrical, you can *pinch it back*—essentially cutting off a portion of new growth, either with your fingers or with scissors. You will almost certainly be loath to do this—after all,

You can encourage your English ivy (page 94) to grow into a bushier plant by trimming any long branches like the one on this plant.

it seems counterintuitive to get rid of some portion of a growing plant—but it actually stimulates even more growth and helps the plant branch out.

Pinching back is particularly useful for vining plants, which can start to look a little scraggly or simply become too long for your taste. You can also use pruning shears to prune branches or stems of larger plants to the same end. Trim these just above a *node*, those little spots along a plant's stem that the leaves grow out of. These plant "haircuts" will take some practice, but you'll become more comfortable with them over time.

BRING OUT YOUR DEAD!

To help keep your plant healthy and looking its best, remove dead and yellow leaves. If you wish, you can also snip off brown leaf tips. Once a flowering plant has finished blooming, remove the dead flowers, a task called *deadheading*. If you find pests, treat them right away. (For more information, see Now What's Wrong? on pages 38–48.) Accept that some extremely infested or diseased plants, especially inexpensive ones, might not be worth saving. Let them go instead of fighting with them.

MOVING ON

Repotting Your Plants

At some point, you will want or need to move a plant to another container. *Repotting* is when you move a plant to a container of the same size, and *up-potting* is when you move your plant to a larger container. (In some parts of the English-speaking world, up-potting is also called *potting on*.)

There are a few key reasons why you might want to repot or up-pot your plant:

> You just purchased a new plant and want to transfer it to a more decorative container.

> The plant has become *pot-bound*, meaning the roots take up so much of the container that it's impossible for the plant to grow any further.

> The soil has become compacted over time as the organic elements in the potting mix break down, making it harder for

the plant's roots to absorb water efficiently. The water might run through so quickly that it's not actually absorbed by the soil.

The best time to do your repotting or up-potting is in the spring and summer. To check whether your plants need to be up-potted, pull them out of their containers and look at the roots. The roots of a pot-bound plant will wrap around the soil. Sometimes the roots will even grow out of the drainage holes.

The tightly packed roots of this aloe vera (page 64) indicate that this plant has outgrown its current home.

Size matters when it comes to pots. When up-potting, the general rule of thumb is to choose a pot with a diameter about 2 inches (5 cm) larger than the current pot. No matter how tempting it may seem, don't put your plant in a much larger pot. The potting mix will stay too wet, which could set you up for a case of root rot (page 44).

Before repotting your plant, check the roots for signs of damage or disease.

The process of potting or up-potting a plant is quite simple. Pull your plant out of its current pot. It can be useful to manually loosen the roots of a pot-bound plant so that they can more easily expand into the larger container. You can also cut off any dead or mushy roots at this time. Pour some potting mix into the new container. Insert the plant into the new pot, making sure it's not sitting too low. Your goal is to have about an inch (2.5 cm) between the top of the soil line and the top of the pot to give you space for watering; add a little more headspace for a big pot and a little less for a small pot. If necessary, remove the plant and add more soil to the bottom of the pot. Then gently start to add soil evenly along the sides, making sure to keep the plant centered in the pot. Add some more soil to the top of the pot. Press down gently, and give your plant a thorough watering. Done!

What happens if you have a plant that's too heavy or big for repotting? If possible, remove the top 1–2 inches (2.5–5 cm) of potting mix and add fresh potting mix on top.

NOW WHAT'S WRONG?

Common Plant Problems

S
ooner or later, you're going to have a problem with your houseplants. In fact, if you keep plants long enough, I'd wager you will have most, if not all, of these problems at one time or another. Try to look at it as an opportunity to learn more about plant care. You'll become a better gardener in the process. Even the experts are not immune to the problems the rest of us mere mortals deal with. This list includes the most common issues you might encounter, not every possible thing that could go wrong with your plants.

I prefer natural, organic solutions to plant problems, rather than spraying something that might be toxic inside the home. Two products to look into are horticultural oils and insecticidal soap, neither of which is chemical-based. Horticultural oils can be petroleum- or plant-based and work by suffocating the pests. For that reason, they need to be sprayed directly on your plant's troublemakers. Insecticidal soap also needs to come in direct contact with the pests. This product is indeed a kind of soap and is thought to work by breaking down or disrupting the membranes of the insects. Rubbing alcohol, soapy water (use a mild dish soap with no antibacterial agents added), and plain water often help as well.

It's easier to take care of a problem in the earlier stages. If the problem is extensive and the plant inexpensive, it's sometimes better to get rid of the plant. It's (usually) not the end of the world. After a severe infestation takes hold, it's often game over.

PESTS

Healthy plants will be less likely to fall prey to insects than stressed plants, so good plant care is your most important weapon against them. These insects can also hitchhike their way into your home on new plants, so check for signs of infestation before you bring a new friend home. You can even quarantine new plants in a separate area until you are sure they are pest free and ready to join the rest of the gang.

In addition to feasting on your plant, many of these insects release a sweet, sticky, shiny substance benignly called *honeydew*. The honeydew can lead to a second problem, sooty mold (page 45), which turns leaves black. Honeydew can also attract ants.

Aphids

You might get excited when your plant grows new leaves or shoots; unfortunately, aphids do too! These teensy pests, which come in a range of colors,

including red, pink, and green, make a beeline for new growth, causing leaves to curl under.

> FIRST AID: Remove aphids by washing the leaves with regular or soapy water. You can also use horticultural oil, which will suffocate them, or insecticidal soap. Avoid using horticultural oil on ferns and other delicate plants unless the label says it's safe to do so. If parts of the plant are heavily infested, remove the leaves in those sections.

Mealybugs

Hard to eliminate and unfortunately quite common, mealybugs are covered in a white, cottony fluff, making them easy to identify. These pests can be found on the underside of leaves and on stems, and they suck plant sap and weaken your plant, slowly killing it. There's another kind of mealybug that goes after the roots of plants. You won't see it unless you remove the plant from its pot. Root mealybugs look like grains of rice and can cause the plant to wilt. (Don't confuse them with the balls of perlite that help provide drainage to your potting mix.)

> FIRST AID: To eliminate leaf mealybugs, apply rubbing alcohol to them with a cotton swab. You can also use a horticultural oil or insecticidal soap according to the manufacturer's instructions. Avoid using horticultural oil on ferns and other delicate plants unless the label says it's safe to do so.

If you have root mealybugs, you've got a real headache. You have to remove all the potting mix that the plant is in, wash off any that remains, and repot your plant in a new container with fresh potting mix. It's sometimes best just to get rid of the plant. You may want to check other plants to make sure they're mealybug free. Disinfect any affected pots after use to prevent the infestation from spreading.

Scale Insects

Brown and oval-shaped, scale insects adhere to the leaves and stems of plants. These little hard-shelled bumps do not move, so you might not even realize that they're sucking the life out of your plant. Like some other  pests, scale insects leave behind honeydew, which can lead to sooty mold. In the case of severe infestations, the honeydew might even coat the surface under your plant, giving you another fun cleanup task.

FIRST AID: You can remove scale by hand, literally rubbing it off with your fingers or a tissue. I prefer, however, to use a damp paper towel or cotton swab dipped in rubbing alcohol. You can also apply horticultural oil. Avoid using horticultural oil on ferns and other delicate plants unless the label says it's safe to do so. Check the plant on a weekly basis to see if the scale has been eradicated. It's easy to miss a few of the little buggers, so a second or third de-scaling is often in order.

Spider Mites

Spider mites often attack in dry conditions. These tiny spiders are too small to see at first, but as the infestation progresses webs will start to form on the leaves and stems. The mites suck on plant sap and can cause leaves to get yellow spots and

sometimes fall off. Leaves may also turn brown. Overall, the plant will appear in poor health. While the webs are a dead giveaway that you've got spider mites, another common way of diagnosing them is to hold a piece of paper under the leaves, then shake the leaves. Any moving specks that fall onto the paper are likely spider mites. When squashed on the paper, spider mites will produce a green streak.

> FIRST AID: To prevent spider mites, pay attention to a plant's watering and humidity requirements, especially in dry conditions. You can use a sharp stream of water from a spray nozzle in the sink to remove some spider mites. A horticultural oil can suffocate them; an insecticidal soap can also kill them.

Whiteflies

Ah, whiteflies. Whenever I try to grow tomatoes on my balcony, all the whiteflies within fifteen miles seem to know and claim squatter's

rights. To identify whiteflies, look under the plant's leaves. You'll see a collection of tiny white insects. To be really sure they're whiteflies, nudge the plant or run your fingers through its foliage. A cloud of these little troublemakers will start flying around. Like scale insects, they also secrete honeydew, which can lead to sooty mold.

FIRST AID: Whiteflies are troublesome because they don't stay in place long enough for you to wipe them away. If you only have a few, act quickly and try to eradicate them as soon as possible. Start by rinsing the leaves off, particularly the undersides. I sometimes use a bit of soapy water and a paper towel to rub off any stragglers. You can reduce their numbers with an insect sticky trap, or also try horticultural oil or insecticidal soap. Once a major infestation has set in, you might have to wave the white(fly) flag and surrender.

PLANT DISEASES

Just as with plant pests, good plant care is key to prevention. Similarly, check any potential purchase for signs of disease.

Gray Mold

Common to African violets and begonias, gray mold is a fungal infection that can develop if plants are crowded or overfertilized, or in the case of African violets, if water collects on their leaves and flowers. Gray mold can cause soft brown or gray spots on leaves and stems as well as rot.

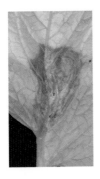

FIRST AID: Completely remove the affected parts of the plant, and improve the conditions that may have led to the gray mold. For example, you may need to reduce the frequency of watering, avoid getting the foliage wet (try bottom watering, page 12), give the plant more space or air circulation, or stop fertilizing. Discard a heavily affected plant, as the problem can spread quickly.

Root Rot

Sometimes called *crown rot*, root rot is a common problem triggered by overwatering, insufficient drainage, and putting a plant in a pot too big for its size. It is not the overwatering itself that causes the rot; instead, the  excess moisture in the soil allows a fungus to set in and attack the roots. In severe cases, the entire plant may collapse.

FIRST AID: Prevention is the best medicine when it comes to root rot. Once it sets in, the treatment is a bit hair-raising, and there's no guarantee of success. You need to remove the plant from its pot, shake off all the potting mix, and cut off any brown, black, or soft roots. Disinfect your pots and any tools you work with so the fungus doesn't spread. Repot in fresh, well-draining potting mix and cross your fingers. If you're already starting to despair, no one will call you a fair-weather friend if the plant just happens to disappear.

Powdery Mildew

Powdery mildew looks like, well, powder, and it can stunt the growth of your plant. It's hard to eradicate, though you can remove any parts of the plant that are affected. It's common to African violets, jade plants, ivy, kalanchoe, and begonias.

FIRST AID: Remove the affected parts of the plant, and improve air circulation around your plants. If the problem is widespread, it might be time to bid the plant a fond farewell.

Sooty Mold

As the name suggests, this fungus looks like soot on plant leaves. It grows on the honeydew (see page 39) secreted by plant pests and blocks light from reaching the plant's leaves. If you have sooty mold, you actually have two problems: the mold itself, and some kind of infestation, which needs to be identified.

FIRST AID: Wash sooty mold off of the leaves with water or a damp cloth. Identify and treat the underlying infestation by referring to the earlier pests section.

ENVIRONMENTAL ISSUES

Environmental problems can be a little tricky to diagnose, because the symptoms that the plant displays for many of them can be similar. The key to overcoming them is to evaluate the growing conditions and the plant's requirements.

Overwatering

Plants that have been either overwatered or underwatered can droop, so to determine which is the problem, stick your finger in the soil. The leaves of overwatered plants can also turn yellow or light green.

> FIRST AID: Put down the watering can! Let the soil dry out, following the watering requirements for each plant.

Underwatering

An underwatered plant may droop, and its leaves may wilt or turn dry and brown and fall off. The soil of an underwatered plant may pull away from the sides of the pot so that when you do water, the water runs quickly around the soil without being absorbed by it.

FIRST AID: Check to make sure the soil really is dry. If it no longer absorbs water, you can soak the pot for 20–30 minutes to rehydrate the soil. If this doesn't work, you can repot your plant in fresh potting mix. Going forward, provide your plant with a more consistent watering schedule.

Sunburn

Just like you, your plant pals can get sunburn! Leaves can turn pale or get brown or black spots. Succulents that are receiving too much light may turn an orangeish or reddish hue.

FIRST AID: Move the plant to a less sunny location. If it's a plant that can tolerate bright light or full sun, acclimate it gradually to the new conditions over a few weeks. See More Sun, Please! on page 8.

Lack of Light

A plant that isn't getting enough light may look leggy (long and slender with weak growth), drop its leaves, or lean toward the light.

FIRST AID: Move the plant to a sunnier spot, but do so gradually. See More Sun, Please! on page 8. You can also rotate the plant a quarter-turn whenever you water it to promote even growth.

Insufficient Humidity

Brown leaf tips, curling leaves, crispy leaves, dropped leaves, and spider mites (page 42) can all be indications that your plant isn't getting enough humidity.

> FIRST AID: Make sure that your plant is getting enough water. If available, move it to a more humid location. Add a pebble tray (page 18), run a humidifier, or group it with other plants.

No Flowers

The causes for lack of flowering can vary, but it's usually due to insufficient light. Some plants, such as the Christmas cactus and the amaryllis, also need a period of cooler temperatures to produce flowers.

> FIRST AID: Move the plant to a location where it gets more light. Review the growing requirements to see if a cool period is needed.

PROPAGATION

Or, How to Make More Plant Friends

A s a beginner, you may ask yourself why you would ever want to propagate, or reproduce, your plants. It seems like something that's best left to experts, but for many plants, however, the task is simpler and more useful than you may think. In fact, some plants will do the work for you.

So, why make more plants? First, some plants only last so long. Once these plants start their decline, you can propagate them as a way of starting over. Second, if you have a plant that you really love, you can create a sort of "backup" in case something happens to the main plant. Similarly, if a plant suffers neglect or other damage, you can often salvage it by taking a cutting of the healthy remaining part, propagating it using the method for that particular species, and discarding the parent plant. Next, propagation allows you to increase your plant collection without spending more money, and you can swap plant babies with plant-loving friends. Finally, you might propagate just for the excitement of being able to get a plant to reproduce. I've selected three relatively easy propagation techniques, the first of which doesn't really require you to do anything but wait.

PLANT BABIES

Many plants, especially succulents, produce offsets, or pups. These plant babies will form in the same pot as the parent plant. Once they are at least a third the size of the main plant, you can use a sharp, clean knife and sever the connection between the two plants. Pot the baby up separately. Some plants, like air plants and bromeliads, will produce offsets after flowering, and the main plant will eventually die after it has reproduced.

Cultivate a small army of spider plants by potting up the parent plant's babies, also known as plantlets.

A spider plant will send out baby plantlets on long stems. While the plantlets are still attached to the main plant, you can place these plantlets in small pots of potting mix. In time, the plantlets will take root, and you can sever the "cord" between them and the parent plant.

LEAF CUTTING

Succulents are probably among the easiest plants to propagate. This can be done by removing a single leaf, called a *leaf cutting*. Take this leaf and let it dry out for a few days. The spot where it was removed from the plant will form a callus. Once the callus has formed, you can either place the calloused leaf tip in water or lay the leaf on damp potting mix and wait until roots appear. Keep the soil only slightly damp to avoid rot. In time, a single leaf will grow roots and an entirely new baby plant at its base. Other plants, like African violets, can be propagated by rooting a leaf in a small pot and covering the leaf with a plastic bag or glass.

STEM-TIP CUTTING

Plants like pothos and Chinese money plant can reproduce if you take a cutting of the plant's stem and leaves and place it in water, where it will eventually grow roots. The stem cutting itself can be quite attractive and displayed decoratively. (You can use a bud vase, empty water bottle, or even a test tube.) Cut a section of the stem at least 5 inches (12.5 cm) long just below a node, the spot that the leaves grow

out of. Remove all but the top few leaves and any flowers. None of the leaves should be in the water, or they'll rot. Put the stem in water and place the cutting in bright indirect light. In a few weeks (sometimes longer), new roots should sprout. Once the roots are a few inches long, you can pot up the new plant.

When you place a stem cutting into water, roots will begin to form, creating an entirely new plant.

A FEW FINAL POINTERS AND REMINDERS BEFORE YOU BEGIN

1. START SMALL. Carefully select a few plants and learn about them. It might be tempting to go big, but if you go out and buy a lot of plants at once, you are more likely to get overwhelmed and make success more difficult once the initial excitement wears off.

2. ASSESS YOUR SPACE. Choose your plants based on your conditions at home, not impulse. And if you do choose based on impulse, give yourself a break if it doesn't work out!

3. BE CONSISTENT. If you give regular attention to your plants, they'll give back to you too.

4. DON'T LOSE HEART IF SUCCESS ISN'T INSTANT. Gardening takes practice. You will learn, observe, and develop an instinct. When (not if) a plant fails, learn from it, let it go, and don't look back.

5. LEARN THE NAMES OF THE PLANTS YOU OWN. Even though it sounds obvious, take the time to learn the name of the plants you buy, so that you can research how best to take care of them.

6. READ THIS BOOK . . . If you take the time to learn the basics, you'll already be miles ahead of where you were before.

7. . . . **BUT DON'T STOP HERE!** Read other books on houseplants to increase your knowledge. Look for resources in your community, such as societies for particular kinds of plants (e.g., orchids and African violets), and seek help from your local extension office. Find a reliable local garden center or nursery. Connect in person or online with other plant lovers who are ready and willing to share their expertise and plant pics with you. As a bonus, you'll get lots of inspiring ideas for how to display plants and will come to know different varieties that you might not have heard of otherwise. You may even find a group of friends willing to do plant swaps and share cuttings with you. Above all, have fun and spread the plant love!

GREATEST HOUSEPLANT HITS

Need help getting started? Use these lists as a shortcut to finding plants to suit your needs. Each list is not exhaustive, of course, but will serve as a quick guide. Check out the plant profiles for more ideas!

Plants That Make a Statement

These floor plants will set the tone in a room.

1. TREE PHILODENDRON (page 154)
2. MONSTERA (page 118)
3. PARLOR PALM (page 132)
4. BRAIDED MONEY TREE (page 76)
5. FIDDLE-LEAF FIG (page 96)

Hanging Plants

A hanging plant can help balance out a plant collection. They're a great choice for small spaces where you want to maximize impact. You can, of course, hang them from a sturdy hook in the ceiling, but you can also attach a planter to the wall or place trailing plants on shelves.

1. BOSTON FERN (page 74)
2. POTHOS (page 140)
3. STAGHORN FERN (page 152)
4. INCH PLANT (page 106)
5. HEART-LEAF PHILODENDRON (page 103)

Great Winter Blooms

Want to add some color to winter? Pick up one of these plants for a burst of blooms.

1. FLAMING KATY (page 98)
2. MOTH ORCHID (page 121)
3. AMARYLLIS (page 66)
4. AFRICAN VIOLET (page 60)
5. CHRISTMAS CACTUS (page 84)

Pet-Friendly Plants

These plants make the American Society for the Prevention of Cruelty of Animals's (ASPCA) list of non-toxic choices for cats and dogs. The ASPCA does note that ingesting any plant material could still cause your pet some gastrointestinal problems, so the best strategy is to teach your pets that plants are a no-go zone.

– Cats –

1. AFRICAN VIOLET (page 60)
2. REX BEGONIA (page 144)
3. BOSTON FERN (page 74)
4. DONKEY'S TAIL (page 88)
5. PARLOR PALM (page 132)

– Dogs –

1. CHRISTMAS CACTUS (page 84)
2. ZEBRA PLANT (page 162)
3. WATERMELON PEPEROMIA (page 158)
4. PONYTAIL PALM (page 138)
5. NERVE PLANT (page 124)

Small-Space Plants

Who says you need a lot of space to have houseplants? These plants may be small in size, but they make a big impact.

1. MINIATURE AFRICAN VIOLETS (page 60)
2. ECHEVERIA (page 92)
3. ZEBRA PLANT (page 162)
4. AIR PLANTS (page 62)
5. CHINESE MONEY PLANT (page 82)

Air-Purifying Plants

Looking to clear the air? These plants should be at the top of your list.

1. **PEACE LILY** (page 134)
2. **POTHOS** (page 140)
3. **RUBBER PLANT** (page 146)
4. **SPIDER PLANT** (page 150)
5. **SNAKE PLANT** (page 148)

Beginner-Friendly Plants

While most of the plants in this book are quite easy to grow, if you're looking for the finalists for the easiest-plant award, here they are.

1. **ALOE VERA** (page 64)
2. **HEART-LEAF PHILODENDRON** (page 103)
3. **POTHOS** (page 140)
4. **SNAKE PLANT** (page 148)
5. **CHINESE EVERGREEN** (page 80)

Low-Light Plants

Don't fret if you don't have a lot of light. While you'll want to be a bit more discerning when choosing your plants, you'll find plenty that will tolerate less-than-optimal light conditions.

1. **CAST IRON PLANT** (page 78)
2. **HEART-LEAF PHILODENDRON** (page 103)
3. **PARLOR PALM** (page 132)
4. **POTHOS** (page 140)
5. **ZZ PLANT** (page 164)

Low-Humidity Plants

Who needs a pebble tray or a humidifier when you've got these plants? If you don't want to be bothered with worrying about whether your Boston fern's fronds are going to turn brown and crispy, start with plants that don't need a lot of humidity. (Hint: A succulent or desert cactus is always a good choice if your home is dry.)

1. CAST IRON PLANT (page 78)
2. CHINESE EVERGREEN (page 80)
3. PONYTAIL PALM (page 138)
4. DONKEY'S TAIL (page 88)
5. JADE PLANT (page 108)

Plants for Nurturers

These plants will benefit from any extra attentiveness you can give, whether through grooming, high humidity, or careful watering.

1. AFRICAN VIOLET (page 60)
2. BOSTON FERN (page 74)
3. NERVE PLANT (page 124)
4. POLKA DOT PLANT (page 136)
5. REX BEGONIA (page 144)

Plants for Frequent Travelers

Not all who wander are lost when it comes to houseplants. Plenty of plants don't mind skipping a watering and can survive a week or two without your attention. Just make sure you give your plants a good watering and grooming right before you leave.

1. ZEBRA PLANT (page 162)
2. ZZ PLANT (page 164)
3. MOTH ORCHID (page 121)
4. SNAKE PLANT (page 148)
5. GOLDEN BARREL CACTUS (page 101)

Part Two

PLANT PROFILES

AFRICAN VIOLET *(SAINTPAULIA)*

Arguably the most popular houseplant, the African violet has pleasantly fuzzy leaves and perky flowers. You may have had good luck with these plants, or you may remember them being a tad difficult—they don't like cold temperatures or water on their leaves. Admittedly, they have had a reputation for being a little frumpy, yet if the idea of having a continuously flowering plant appeals to you, it's time to have another look. There are now so many beautiful varieties that you're virtually certain to find one you like, and once you understand them a bit better, you'll find them quite easy to care for. You might just find yourself seeing African violets in a whole new way.

How to Make Friends with African Violets

» SOIL

Use African violet potting mix, which is light, porous, and well-draining.

≫ LIGHT

African violets like bright indirect light, such as that from an eastern or western window. Avoid direct sun. Rotate the plants a quarter-turn regularly for even growth.

≫ WATER AND HUMIDITY

The general recommendation is to water African violets from the bottom (page 12), to avoid splashing their delicate leaves, but if you have a watering can with a narrow spout, you can also water from the top, being careful to avoid the leaves. Don't let the plant dry out entirely. In dry conditions, use a pebble tray (page 18) to increase humidity.

≫ TEMPERATURE

Room temperatures of 60–85°F (16–30°C) are suitable for African violets. Cold temperatures can stunt their growth.

≫ SIZE

African violets range in size from dollhouse-sized microminiatures (1 inch/2.5 cm in diameter) to large plants (16 inches/40 cm in diameter). Most plants are suitable for a windowsill, shelf, or table.

≫ BUGS AND DISEASE

Watch out for mealybugs (page 40) and gray mold (page 43).

≫ OTHER TIPS

Because they have shallow root systems, African violets prefer small pots. African violets are easy to propagate from a leaf cutting (page 50). Remove dead flowers; not only will the plant look nicer and stay healthier, but it can promote additional blooming.

AIR PLANTS *(TILLANDSIA)*

You may have seen these intriguing plants placed under glass cloches, hanging in geometric brass ornaments, nestled on bookshelves, or tucked in terrariums. They look like living works of abstract art or alien creatures, depending on your viewpoint. A kind of bromeliad, tillandsias—with a few exceptions—don't grow in soil, hence the name "air plant." Some common species to keep for houseplants include *T. ionantha*, *T. xerographica*, and *T. caput-medusae*, which does look a little like the head of Medusa.

Air plants are quite easy to care for. They like bright light, and the more sun they get, the more water they need. The plant will tell you when it's thirsty: the leaves will curl back more than usual and the tips may become dry. Just follow the unique watering regime listed below to keep your air plants happy and healthy.

How to Make Friends with Air Plants

≫ SOIL

None! Air plants are a perfect choice for someone who doesn't like to get their hands dirty.

» LIGHT

Air plants like bright light, such as that from an eastern or western window. Several cultivars, including *T. xerographica*, can tolerate more light.

» WATER AND HUMIDITY

Soak your air plants in filtered water for about 30 minutes once a week. Allow them to dry completely, upside down, before returning them to their display; otherwise, they might rot. You can also mist on a daily basis, or combine regular misting with occasional soaking.

» TEMPERATURE

Air plants are suitable for room temperatures of 50–90°F (10–32°C).

» SIZE

Air plants are often small enough to fit in the palm of your hand. Some species, however, can reach several feet in diameter.

» BUGS AND DISEASE

Bugs and disease are uncommon with air plants, but watch out for signs of overwatering or underwatering.

» OTHER TIPS

Air plants like good air circulation. If your home doesn't have good airflow, consider setting up a small fan nearby.

After your plant flowers, it will produce several pups, or offsets. You can remove these babies and grow them as individual plants once they are at least a third the size of the main plant.

ALOE VERA

Aloe vera should be part of any houseplant lover's collection. With pleasantly plump and fleshy leaves, it's one of the best houseplants for beginners because it's incredibly easy to care for. Plus it looks great in almost any kind of pot—from simple terra-cotta to sleek and shiny alternatives.

Not only is aloe simply a great plant to have around, it has a practical use too. Often called the *burn plant* or *first-aid plant*, it is used to treat minor skin irritations and burns. If you've ever bought that artificially colored green sunburn gel at the drugstore, you've used some form of it, but this is the real deal.

If you're not sold on aloe yet, here's one more bonus: it has babies. They're called pups, and once the pups get big enough, you can put them in adorable little pots and give them to all your friends.

How to Make Friends with Aloes

» **SOIL**

Use a cactus and succulent mix.

» **LIGHT**

Aloe plants like the sun, so put them in a south- or west-facing window, if possible.

» **WATER AND HUMIDITY**

The best way to kill your aloe plant is to overwater it. Unless that's your goal, wait until the top inch (2.5 cm) of the soil feels dry. Reduce watering frequency in winter. Aloe doesn't have any special humidity requirement.

» **TEMPERATURE**

Aloe can adapt to room temperatures above 50°F (10°C).

» **SIZE**

Most aloe vera plants are between 1 and 2 feet (30–60 cm) tall.

» **BUGS AND DISEASE**

Problems are uncommon, but watch out for mealybugs (page 40) and scale (page 41).

» **OTHER TIPS**

Aloe pups will appear at the base of the plant. You can cut these out with a clean, sharp knife. This seems a bit brutal, but that "wound" will callus over. Set the pups aside for a few days until this happens, and then plant them in small pots of cactus potting mix. Don't water until they start growing roots. (To check if roots have formed, tug very gently on the pup.)

AMARYLLIS (HIPPEASTRUM)

With its strappy leaves and riot of large blooms, a flowering amaryllis can feel like a godsend in the darkest days of winter. These plants are particularly popular around the holidays, and that's often the best time to pick up a kit. A kit can be the easiest way to get started because it contains everything you need: an amaryllis bulb, potting mix, and a pot. Five to eight weeks after planting your bulb, you'll be basking in the glow of the gorgeous flowers. The flowers come in a range of colors,

including pink, white, red, and orange, and there are also beautiful variegated versions. Because an amaryllis grows from a bulb, it can rebloom the following year after a period of dormancy. If you'd like to give reblooming a try, check out the instructions on page 68, though no one will be the wiser if you simply pick up a new plant when you want your flower fix.

How to Make Friends with Amaryllises

≫ SOIL

Use an all-purpose potting mix.

≫ LIGHT

Amaryllis likes bright light, such as that from an eastern window.

≫ WATER AND HUMIDITY

Keep the soil moderately moist, but never soggy. As the dormancy period approaches, reduce watering frequency. During dormancy, watering is not necessary, but you can sprinkle the soil occasionally with water. Average room humidity is fine; in particularly dry rooms, use a pebble tray (page 18).

≫ TEMPERATURES

Room temperatures of 60–80°F (16–27°C) are suitable for amaryllises.

≫ SIZE

Amaryllises are medium-sized plants that can be displayed on a table, plant stand, or even a wide windowsill. The flower stalks can be 2 feet (60 cm) high.

≫ BUGS AND DISEASE

Watch out for mealybugs (page 40).

If you are planting your own amaryllis bulb, there's a special trick to it. Use a small pot that is only slightly bigger than the bulb, and leave the top third to top half of the bulb exposed.

If you'd like to get your plant to rebloom, cut off the flower stalk once the flowers die and continue to care for the plant, giving it plenty of sun. (You can even put it outdoors if you'd like.) Come autumn, begin reducing the frequency of watering until you eventually stop watering it entirely. This will help the plant go dormant. Remove the now-dead leaves, and put the pot in a cool, dark place (around 50°F/10°C) for a few months. About six weeks before you want the plant to bloom, take it out of its resting place and give it light and water. Once the flower spike appears, rotate the plant regularly so that the stalk grows evenly, adding a bamboo stake to prop it up if necessary. Get ready for greatness.

ARROWHEAD PLANT (SYNGONIUM PODOPHYLLUM)

The arrowhead plant is a classic for a reason. It has beautiful, usually variegated, arrow-shaped foliage, and it's a breeze to grow, as long as the room isn't super-dry. If you want a plant that doesn't need a ton of light, it's a great choice. It will tolerate a north-facing window, though it would be happier in a slightly brighter place where the light is filtered and indirect. While your plant may be compact for quite some time, it will eventually become a vine, which will look lovely in a tall container or hanging basket or trained up a pole. You can also keep it pruned back so that it looks bushier instead of vine-like.

How to Make Friends with Arrowhead Plants

» SOIL

Use an all-purpose potting mix.

» LIGHT

Arrowhead plants like bright indirect light, such as that from an
east window. A bright north window will work as well. Avoid
direct sun, which is too strong for this plant and may cause the
leaves to turn pale.

» WATER AND HUMIDITY

Keep the potting mix consistently moist, watering when the top
half-inch (1 cm) of the soil feels dry. Reduce the frequency of
watering in the winter. In dry rooms, place the plant on a pebble
tray (page 18) to raise humidity. For even growth, turn this plant a
quarter-turn every time you water it.

» TEMPERATURE

Room temperatures of 60–80°F (16–27°C) are suitable for
arrowhead plants, though they can tolerate somewhat warmer
temperatures as well.

» SIZE

Before an arrowhead plant starts to become a vine, it is around
12–15 inches (30.5–38 cm) tall. The vines can be several feet long.

» BUGS AND DISEASE

Watch out for root rot (page 44), scale (page 41), and mealybugs
(page 40).

BIRD'S NEST FERN
(ASPLENIUM NIDUS)

If you like unusual plants, check out the bird's nest fern. The first time I saw one, I was immediately drawn to its rippled, almost crimped glossy leaves. This fern is not delicate or lacy like other types but almost looks like wiry ribbon. New fronds, shaped like tiny eggs, unfurl from the nest-like center of the plant. Other than the plant's attractive appearance, the best thing about a bird's nest fern is that it's easy to care for and can tolerate lower light levels quite well. The only real trick to

remember is to water around the sides of the plant, not into the center where the new leaves are forming. If water pools in the little "nest," the plant could rot.

How to Make Friends with Bird's Nest Ferns

>> SOIL

Use an all-purpose potting mix that drains very well.

>> LIGHT

A bird's nest fern likes moderate light, such as that from a northern or eastern window. Keep it out of direct sunlight.

>> WATER AND HUMIDITY

Keep the soil consistently moist but not soggy, watering when the top half-inch (1 cm) of the soil feels dry. While you should really try not to let it dry out between waterings, it can survive occasional neglect, especially in more humid climates. Reduce watering frequency somewhat in winter. This plant likes humidity, so add a pebble tray (page 18) in dry conditions, or place the plant in a bathroom.

>> TEMPERATURE

Room temperatures of 55–80°F (13–27°C) are suitable for a bird's nest fern.

>> SIZE

New plants may be 6–8 inches (15–20 cm) tall. Over time, a happy plant may grow to 18–24 inches (46–60 cm) or even taller.

» BUGS AND DISEASE

Watch out for scale (page 41). If you discover an infestation, remove the bugs by hand or with a damp paper towel. If using an insecticidal soap or horticultural oil, check the label to see if it is safe to use with ferns, as their fronds are sensitive.

» OTHER TIPS

If you notice the older outer leaves turning brown, don't worry; it's normal! Just remove them as part of your regular plant grooming. For leaves that don't come off easily when you pull lightly, feel free to snip them off. If other leaves turn brown, however, it may be a sign that the plant is not getting enough water or humidity, or that you've treated the plant with a chemical it didn't like.

BOSTON FERN
(NEPHROLEPIS EXALTATA 'BOSTONIENSIS')

What comes to mind when you hear the word *fern*? Chances are that you think of a bushy plant that resembles a green lion's mane. The Boston fern fits that description perfectly. This fern can add a feeling of lushness to any room, particularly if you hang it in a basket or place it on a pedestal so that its long fronds can dangle. Many ferns have a reputation for being a bit tricky to grow (we're talking about you, maidenhair fern!), but it's fairly easy to keep a Boston fern healthy as long as you can provide it with enough humidity and water. In fact, it can be a great choice for those who like to fuss over their plants. For a less hands-on fern, check out the bird's nest fern (page 71).

How to Make Friends with Boston Ferns

>> SOIL

Use an all-purpose potting mix.

>> LIGHT

Medium indirect light is best, such as that from an eastern
window. If the light is filtered through a sheer curtain, you can
also place a fern near a southern or western window.

>> WATER AND HUMIDITY

Keep the soil consistently moist but not soggy, watering when
the surface of the soil feels dry. Do not let it dry out; it may drop
leaves. Mist regularly or use a pebble tray (page 18) to satisfy this
plant's love of humidity. A bathroom with good light is an excellent
spot for a Boston fern.

>> TEMPERATURE

Room temperatures of 55–80°F (13–27°C) are suitable.

>> SIZE

When well-grown, these ferns can get quite large, 2–3 feet
(60–90 cm) tall and wide.

>> BUGS AND DISEASE

Watch out for scale (page 41), spider mites (page 42), and
mealybugs (page 40). If using an insecticidal soap or horticultural
oil, check the label to see if it is safe to use with ferns.

>> OTHER TIPS

If the lower fronds are faded or brown, give the plant a little
haircut and trim them off.

BRAIDED MONEY TREE (PACHIRA AQUATICA)

In feng shui, the braided money tree is thought to bring good luck and wealth. Regardless of whether you are buying the plant with an eye to improving your fortune, this beautiful and popular tree is worth a look. It's easy to grow and is typically sold with a braided trunk. The leaflets grow in a palmate arrangement (meaning that they look like the fingers of an open hand) around a central stem.

As the botanical name suggests, these tropical plants grow in wet conditions, such as swamps and riverbanks, but it's not necessary to flood them with water. Just keep the soil moderately moist. Because the plant grows best in bright indirect light, you have a bit of flexibility as to where you can place it, which is nice for a larger floor plant like this one. You may also see these plants labeled Guiana or Malabar chestnut.

How to Make Friends with Braided Money Trees

» SOIL

Use a well-draining all-purpose potting mix. You can improve drainage by adding sand to the mix.

>> LIGHT

A braided money tree likes bright indirect light, such as that from an eastern or western window. If the plant is getting too much sun, the leaves may turn yellow or fall off.

>> WATER AND HUMIDITY

Water when the top inch or two (2.5–5 cm) of the soil feels dry. If the plant drops leaves, it can be an indication that the soil is too dry, but check before giving it more water, as this can also happen if the plant has been recently moved or is receiving too much light. Reduce the frequency of watering in the winter. This plant likes a moderate amount of humidity. If your home is dry, consider putting it on a pebble tray (page 18).

>> TEMPERATURE

Room temperatures of 60–80°F (16–27°C) are suitable for a braided money tree.

>> SIZE

This floor plant can eventually grow up to 8 feet (2.4 m) tall in the home.

>> BUGS AND DISEASE

Watch out for scale (page 41), spider mites (page 42), and root rot (page 44).

>> OTHER TIPS

This plant can be sensitive to drafts, so place it away from heating and air-conditioning vents and drafty windows.

CAST IRON PLANT (ASPIDISTRA ELATIOR)

The cast iron plant's name is a testament to its indestructibility. Aspidistra's popularity dates back to Victorian times, when it became a fixture in parlors and bars because it could tolerate low light, poor air quality, a wide range of temperatures, and general neglect. This plant was so iconic that George Orwell wrote a 1936 novel titled *Keep the Aspidistra Flying*, a critique of middle-class British society.

If your home only has indirect light, you should run, not walk, to the nearest nursery and get yourself an aspidistra. Its strappy green leaves and elegantly long stems can look absolutely stunning, especially if you play up its chic minimalism with the right pot. It grows slowly, so it doesn't require regular repotting, and it suffers almost every abuse but overwatering and direct sun without complaint. It also now comes in many interesting cultivars if plain green isn't your thing.

How to Make Friends with Cast Iron Plants

>> SOIL

Use an all-purpose potting mix.

>> LIGHT

A cast iron plant likes low light, making it an excellent choice for those who only have north-facing windows. Avoid direct sun.

>> WATER AND HUMIDITY

Water your cast iron plant when the soil is dry to a depth of 1½–2 inches (4–5 cm). If you forget to water it occasionally, it should suffer no ill effects; on the other hand, it hates to be overwatered. Humidity or lack thereof is not an issue.

>> TEMPERATURE

Unsurprisingly, the cast iron plant can tolerate a wide range of room temperatures (50–85°F/10–30°C).

>> SIZE

A cast iron plant is typically about 2 feet (60 cm) tall, suitable for a plant stand, table, or even the floor.

>> BUGS AND DISEASE

Watch for scale (page 41) and root rot (page 44).

>> OTHER TIPS

Clean the long leaves periodically with a damp cloth to remove dust.

Propagation is quite easy: divide the plant and pot up the divisions separately.

CHINESE EVERGREEN (AGLAONEMA)

While no living plant is entirely foolproof, a Chinese evergreen should be on any list of the easiest houseplants for beginners. It tolerates low light, doesn't mind some watering neglect, isn't prone to pests or diseases, and helps purify indoor air. It's a win-win-win-win. The most popular and hard-to-kill varieties include 'Emerald Beauty' and 'Silver Queen', and they have pretty variegated leaves in shades of green and white. Newer varieties come in more brightly colored reds and pinks, and they'd prefer a bit more light, if you please. Regardless of the type you choose, this is one steadfast plant friend.

If you're looking for a medium-sized, slow-growing plant that will just sit there and do its thing, a Chinese evergreen is for you. Set it on a corner table to brighten up any room.

How to Make Friends
with Chinese Evergreens

» **SOIL**

Use an all-purpose potting mix.

» **LIGHT**

Most Chinese evergreen plants do well in low light, such as that
from a north-facing window. If you have a brightly colored variety,
put it in a brighter location, such as an east-facing window.

» **WATER AND HUMIDITY**

Water your Chinese evergreen when the top inch or two
(2.5–5 cm) of the soil feels dry. It will tolerate an occasional
missed watering. Humidity levels are not a particular issue
with this plant.

» **TEMPERATURE**

Room temperatures of 60–80°F (16–27°C) are suitable for this
plant.

» **SIZE**

The size of Chinese evergreens varies by species, but many plants
are 1–2 feet (30–60 cm) tall.

» **BUGS AND DISEASE**

Watch out for scale (page 41).

» **OTHER TIPS**

The leaves tend to collect dust, so wipe them down periodically
with a damp cloth.

CHINESE MONEY PLANT (PILEA PEPEROMIOIDES)

Originally from the Yunnan province of China, the Chinese money plant (also known as UFO plant and pancake plant) has become something of a darling in recent years, fueled perhaps in part by its hard-to-find status. Doubtless, part of the charm is also its cute appearance—the flat leaves recall lily pads (or coins, of course) and seem to almost bounce on the stems. It's also easy to grow and propagate; in fact, it produces a lot of its own babies, and they're often passed from friend to friend.

Keep in mind that "pilea" is a genus containing hundreds of species. When you go into a nursery make sure to specify which one you're referring to.

How to Make Friends
with Chinese Money Plants

» SOIL

Use a well-draining, all-purpose potting mix, or a cactus and succulent mix.

» LIGHT

The Chinese money plant likes moderate light, such as that from an eastern window. Rotate the plant a quarter-turn weekly for even growth. Keep it out of direct sun, which could cause sunburn.

» WATER AND HUMIDITY

Water when the top half inch to inch (1–2.5 cm) of the soil feels dry. These plants like some humidity, so consider using a pebble tray (page 18) in dry conditions.

» TEMPERATURE

Room temperatures of 60–80°F (16–27°C) are suitable for this plant.

» SIZE

This relatively small plant is often up to 12 inches (30 cm) tall, but older, well-cared-for specimens could be much larger. Generally, they're perfect for a windowsill or other small space.

» BUGS AND DISEASE

Watch out for powdery mildew (page 45) or root rot (page 44).

» OTHER TIPS

The flat leaves tend to accumulate dust. Clean with a damp cloth.

The plant will produce babies at the base of its stem, which you can cut off with a sharp, sterile knife, and pot up. You can also propagate it via stem cuttings (page 51).

CHRISTMAS CACTUS (SCHLUMBERGERA BUCKLEYI)

Who doesn't need a few gorgeous flowers on a gray winter day? While you might not be able to time your Christmas cactus to bloom exactly on December 25, it will flower reliably in winter with large, brightly colored blossoms every year. These plants also live a long time. Case in point: My mother's Christmas cactus is easily more than 40 years old. Once the buds have set, don't move the plant too much, or it may lose some of the flowers. Considered a rainforest cactus, the Christmas cactus does like more water and humidity than your average desert cactus. It doesn't technically have leaves, but flat, segmented stems with spiky edges.

How to Make Friends
with Christmas Cactuses

» SOIL
Use a mixture of all-purpose potting mix and sand.

» LIGHT
A Christmas cactus will be happiest with bright light from an
east- or west-facing window. Without protection, a south-facing
window may be too strong.

» WATER AND HUMIDITY
Water when the top 1–2 inches (2.5–5 cm) of the soil feel dry. Don't
let the soil dry out entirely, but it's better to underwater than risk
root rot. Reduce the frequency of watering in winter. The Christmas
cactus likes humidity. If your home is dry, use a pebble tray (page 18).

» TEMPERATURE
Room temperatures of 55–80°F (13–27°C) are suitable. Keep the
temperature on the cooler end of the range during fall and winter.

» SIZE
Most plants are relatively small, about 1 foot (30 cm) tall and wide.

» BUGS AND DISEASE
Watch out for root rot (page 44).

» OTHER TIPS
For blooms every year after year, make sure the plant is in a cool
room (below 65°F/18°C) when fall arrives and that it experiences
long, dark nights—don't even turn on a light. Christmas cactuses are
easy to propagate. Take a three-segment cutting of the plant stem
and pot it up!

DIEFFENBACHIA
(DIEFFENBACHIA SEGUINE)

Few plants have a common name that references their toxicity, but dieffenbachia, aka dumb cane, does. If the plant is consumed, it can render a person "dumb" and unable to speak because the calcium oxalate crystals contained in the plant can burn the mouth and cause swelling and even paralysis of the vocal cords. Yikes! Unsurprisingly, it's toxic to pets too. (This is a reminder that many houseplants are toxic to humans and animals. Visit the resources section on pages 168–169 for more info. Call a doctor, vet, or poison control center if you suspect that a person or pet has consumed any houseplant.) Toxicity

aside, dieffenbachia is a well-loved, easy-going, and moderately sized houseplant with long leaves in varying shades of white and green. While it will grow best in moderately bright filtered light, it will also do well in indirect light, even the light from a north-facing window.

How to Make Friends with Dieffenbachias

>> SOIL

Use an all-purpose potting mix.

>> LIGHT

Dieffenbachia likes moderately bright indirect or filtered light, such as that from an east or west window. A north-facing window is acceptable as well.

>> WATER AND HUMIDITY

Water when the top inch (2.5 cm) of the soil feels dry. Water less frequently in fall and winter. When watering, rotate it a quarter-turn so that it grows evenly. This plant likes humidity, so if your home is dry, add a pebble tray (page 18) under the plant.

>> TEMPERATURE

This plant prefers room temperatures of 65–80°F (18–27°C).

>> SIZE

Many cultivars are between 1 and 2 feet (30 and 60 cm) tall, although some can be even larger, making this plant ideal for a table or plant stand.

>> BUGS AND DISEASE

Watch out for mealybugs (page 40) and spider mites (page 42).

>> OTHER TIPS

This plant's sap can irritate the skin, so wear gloves when handling it.

If the leaf tips turn brown, the plant might have a buildup of fertilizer salts. Take the plant to the sink and let the water run through the potting mix to flush out excess salts.

DONKEY'S TAIL
(SEDUM MORGANIANUM)

Donkey's tail is a delicate succulent popular for its stunning appearance. Its dusty greenish-blue stems do resemble the tail of a donkey (or long, juicy asparagus tips, if you're me); some other common names for this plant are burro's tail, horse's tail, and lamb's tail. Donkey's tail looks particularly striking potted up in hand-thrown ceramics or hanging baskets so its stems can dangle down. Like many succulents, it's quite easy to care for; the only caveat is that those tiny baby leaves fall off at the slightest nudge, so the less you handle this plant, the better. However, the fallen leaves can be a good thing, because you can start experimenting with propagation. Who would not want more of this plant? The pink flowers, which can appear in summer, are quite lovely as well.

How to Make Friends with Donkey's Tails

≫ **SOIL**

Use a cactus and succulent mix.

≫ **LIGHT**

Donkey's tail likes bright light and can handle some full sun if you acclimate it gradually. An east-facing window would work well, and you can also put it slightly back from a west-facing window. For southern and western exposure, watch the plant carefully for sunburn, and adjust the placement as needed.

≫ **WATER AND HUMIDITY**

Water when the top inch (2.5 cm) or so of the soil feels dry. Water less frequently in winter. No supplemental humidity is needed.

≫ **TEMPERATURE**

Room temperatures of 65–75°F (18–24°C) are ideal for most of the year, but donkey's tail prefers cooler temperatures in winter.

≫ **SIZE**

The size of this plant can vary, depending on how old and well grown it is. When you first buy it, it may easily fit in a small pot placed on a windowsill; but over time, the stems can reach 2–3 feet (30–60 cm), making it a good fit for a hanging basket.

≫ **BUGS AND DISEASE**

Watch out for root rot (page 44), aphids (page 39), and sunburn (page 47).

≫ **OTHER TIPS**

You can make more of this plant with leaf cuttings (page 50).

DWARF UMBRELLA TREE (SCHEFFLERA ARBORICOLA)

The leaves on a dwarf umbrella tree are most intriguing. Leaflets are arranged in a circle around the stem, giving the impression of an umbrella, an octopus (octopus tree being another common name for the plant), or even fingers. This plant is sedate, neat in appearance, and slow-growing. As such, it's not going to turn your home into a jungle. By pruning it, you can keep its bushy and clean look and maintain the height you want.

I have seen artificial versions of this plant in commercial settings, but it's so easy to care for that I wonder why you would even bother. Just give it some moderate light and water it occasionally. The plant does have a tendency toward phototropism, meaning that it will tilt toward the light; give it a quarter-turn once a week to promote even, straight growth.

How to Make Friends
with Dwarf Umbrella Trees

» **SOIL**

Use an all-purpose potting mix.

» **LIGHT**

Dwarf umbrella trees like moderate indirect or filtered light, such as that from an eastern or western window. Avoid direct sun. If you have a variegated umbrella tree that is losing its coloring, it's not getting enough light.

» **WATER AND HUMIDITY**

Let the top inch (2.5 cm) of the soil dry out before watering your plant. Be careful not to overwater it because it could lead to root rot. On the other hand, if you underwater it, the leaves will start to droop. Additional humidity will benefit this plant in dry homes. Mist occasionally and/or use a pebble tray (page 18).

» **TEMPERATURE**

Room temperatures of 60–80°F (16–27°C) are suitable for this plant.

» **SIZE**

This is a floor plant that can eventually reach 3–6 feet (1–1.8 m) tall.

» **BUGS AND DISEASE**

Watch for scale (page 41), spider mites (page 42), and root rot (page 44).

» **OTHER TIPS**

Keep the leaves clean by wiping them occasionally with a damp cloth or giving the plant a shower. It will love the extra humidity!

ECHEVERIA

Sometimes called *hens and chicks*, echeverias are rosette-shaped succulent plants from warm, semi-arid climates. (They are sometimes confused with sempervivums, which can grow in colder conditions—even the Alps.) Once you start checking out all the available echeveria species and cultivars, you might just want them all; they're so darling! Since many of these plants are are small, why not put a selection of your favorites together in a dish garden?

Like most succulents, echeverias are great for beginners. They like warmth, lots of light, and not too much water. In fact, these plants can handle a bit of drought. If you travel a lot or have a tendency to ignore your plants, echeverias are a good choice.

How to Make Friends with Echeverias

» SOIL
Use a cactus and succulent potting mix.

» LIGHT
Put your echeverias in bright light to full sun, such as that from a south- or west-facing window.

» WATER AND HUMIDITY
Allow the top inch (2.5 cm) or so of the soil to dry before watering. Water less often in the fall and winter. Overwatering can lead to root rot. Humidity levels are unimportant.

» TEMPERATURE
Room temperatures of 60–80°F (16–27°C) are suitable for echeverias.

» SIZE
Many of these plants can be quite small, just a few inches across, making them quite perfect for a sunny windowsill.

» BUGS AND DISEASE
Watch out for mealybugs (page 40), root rot (page 44), and sunburn (page 47).

» OTHER TIPS
If your plant sends up random stems from the rosette, don't worry; they're flower stalks! Echeveria are easy to propagate a new plant from a single leaf cutting (page 50).

Looking for a few options? The 'Topsy Turvy' cultivar looks like a gray-green blooming onion. 'Perle von Nurnberg' has beautiful hints of purple and pink. Also check out the popular *E. secunda*, *E. elegans*, and *E. imbricata*.

ENGLISH IVY *(HEDERA HELIX)*

Ahh, ivy. There's something *Secret Garden* romantic about this plant, which inspires images of ivy-covered cottages and walls. Romance aside, those creeping vines can be invasive in some parts of the world and occasionally cause damage to walls, gutters, and siding. Indoors, ivy makes a lovely, resilient houseplant, and there are so many different kinds to choose from. Imagine it trailing down a bookshelf or from a pedestal planter. You can also train it upward or into different shapes. Just trim it as needed to keep it under control and looking sharp.

The other nice thing about ivy is that you don't need a lot of light for it to thrive. If you only have a north-facing window, give ivy a try. It's also a great choice for cooler rooms, but it does enjoy humidity.

How to Make Friends with English Ivy

≫ SOIL

Use an all-purpose potting mix.

» LIGHT

Ivy likes moderate to bright indirect or filtered light, such as that from a northern or eastern window. Variegated ivies will need a bit more light, or the leaves will revert to a solid color. Avoid full sun and direct light.

» WATER AND HUMIDITY

Keep the soil consistently moist but not soggy, allowing the top half inch (1 cm) to dry out before watering. Water less frequently in fall and winter. Maintain high humidity levels in dry conditions with a pebble tray (page 18).

» TEMPERATURE

Cooler room temperatures of 50–70°F (10–21°C) are optimal, but ivy can tolerate temperatures outside this range.

» SIZE

Ivy is a vining or trailing plant that can vary in size.

» BUGS AND DISEASE

This plant is prone to spider mites (page 42). Humidity will help keep them at bay, as will the occasional shower to rinse off the leaves.

» OTHER TIPS

There is one tricky thing to watch out for with ivy: if you overwater it, the leaves can—paradoxically—dry out and turn brown. This can also happen with insufficient humidity.

Ivy can easily be propagated with stem-tip cuttings (page 51).

FIDDLE-LEAF FIG *(FICUS LYRATA)*

The genus *Ficus* encompasses around 900 plant species commonly known as figs. *Ficus carica*, for example, produces the kind of figs you buy at the supermarket—yum! The particular species known as the fiddle-leaf fig is a beautiful plant that has gotten so popular in

recent years that you can't visit a design blog, Instagram account, or furniture store without seeing about twelve of them. These plants can grow bushy or tall and have attractive, large, fiddle-shaped leaves that just seem to add an instant element of calm and coziness to a room. If you give your fiddle-leaf fig bright light (not full sun), it will thrive and your only problem might be how tall it gets.

How to Make Friends with Fiddle-Leaf Figs

» SOIL

Use an all-purpose potting mix.

≫ LIGHT

A fiddle-leaf fig likes bright light, but can be sensitive to too much sun; an east-facing window is ideal. If your tree is growing in a west-facing window, place it back slightly from the window or use a sheer curtain to diffuse the hot afternoon sun.

≫ WATER AND HUMIDITY

Water when the top inch (2.5 cm) of the soil feels dry. Your fig likes regular waterings but doesn't want soggy soil. Water less frequently in winter. In dry rooms or climates, place the plant on a pebble tray (page 18) to increase humidity.

≫ TEMPERATURE

Room temperatures of 60–80°F (16–27°C) are suitable for this plant.

≫ SIZE

A fiddle-leaf fig can reach 6–10 feet (1.8–3 m) in height.

≫ BUGS AND DISEASE

Watch out for spider mites (page 42), aphids (page 39), and scale (page 41).

≫ OTHER TIPS

The large leaves, which grow up to 18 inches (46 cm) long, collect dust. Wipe them down periodically with a soft, damp cloth to keep the plant healthy.

To control the size of this large plant, don't repot it too often. You can prune both the root ball and the branches to limit the plant's size, but do so with care: the fig will bleed a sticky sap that contains latex, which some people are allergic or sensitive to. Wear gardening gloves and consider taking the plant outside for the task, to keep the sap from dripping on the floor.

FLAMING KATY
(KALANCHOE BLOSSFELDIANA)

How can you resist a plant with the evocative name *flaming Katy*? And look at all of those flowers! It could not look more different than the other kalanchoe in this book, the panda plant (page 130). However, both are succulents and have similar needs. Flaming Katy is particularly popular around the holidays for its long-lasting, bright flowers in shades of red, pink, yellow, and orange, but you can buy it and enjoy it at any time of year. While the blooms are the real show, the

scallop-edged leaves are pretty on their own. If you plan to keep it as a foliage plant, cut back the flower stems after they die. If you're lucky, you may have a few stray flower stems pop up here and there without any special effort.

How to Make Friends with Flaming Katy

>> SOIL

Use a cactus and succulent potting mix.

>> LIGHT

Flaming Katy would love full sun from a south-facing window, but you should acclimate the plant gradually to this level of light to avoid sunburned leaves. Bright light (including indirect light) from an east- or west-facing window will work well too.

>> WATER AND HUMIDITY

Water when the top inch (2.5 cm) of the soil feels dry. Water less frequently in winter. Overwatering could cause stem rot, meaning that the stem will turn black and soft and the affected part of the plant will most likely die. This plant doesn't have any specific humidity requirements.

>> TEMPERATURE

Room temperatures of 65–85°F (18–30°C) are suitable for flaming Katy.

>> SIZE

Flaming Katy is a small plant (around 10–12 inches/25–30 cm tall).

>> BUGS AND DISEASE

>> BUGS AND DISEASE

Watch out for stem rot, spider mites (page 42), and mealybugs (page 40).

>> OTHER TIPS

The process for getting the plant to rebloom is not difficult, but it is multi-stepped, involving alternately exposing the plant to sun, controlled temperatures, and long, dark nights for a number of weeks. Then you wait for buds to appear. If you're interested in all the details, there are many web resources or books available that can walk you through all the steps.

Flaming Katy can be easily propagated with stem-tip cuttings (page 51).

GOLDEN BARREL CACTUS (ECHINOCACTUS GRUSONII)

Cactuses can be polarizing. I like the desert style these living sculptures can bring to a home. Others find their prickly nature a bit off-putting—and that's okay too. (They surely don't invite cuddling.) While cactuses have a reputation for being impossible to kill, you can certainly find a way if you give them too much water or not enough light. Think about a cactus's natural environment: it's very dry and sunny, with almost no shade in sight. In general, the best place for a cactus in your home is the brightest spot you can find. Then do very little to it after that.

Shaped like a deeply ribbed pincushion, the golden barrel cactus is one of the most popular to keep as a houseplant. It can live many, many years if well taken care of, but it grows rather slowly.

How to Make Friends with Golden Barrel Cactuses

≫ SOIL
  Use a cactus and succulent mix.

≫ LIGHT

A golden barrel cactus likes the sun. Place it in or near a south-facing window.

≫ WATER AND HUMIDITY

Water when the top 2 inches (5 cm) of the soil feel dry. Water less frequently in winter, when overwatering can easily cause root rot. Cacti are used to dry air, so there's no need to add supplemental humidity.

≫ TEMPERATURE

Room temperatures of 55–75°F (13–24°C) are suitable for this cactus. Temperatures toward the cooler end of this range are beneficial in winter.

≫ SIZE

Although it's possible for this plant to reach several feet in height indoors, it grows very slowly and is generally a small houseplant (around 4–10 inches/10–25 cm).

≫ BUGS AND DISEASE

Watch out for root rot (page 44), mealybugs (page 40), and spider mites (page 42).

≫ OTHER TIPS

How do you repot a cactus other than very carefully? Wear thick gardening gloves and use tongs. You can also remove the cactus with several sheets of newspaper.

HEARTLEAF PHILODENDRON (PHILODENDRON HEDERACEUM, SYN. PHILODENDRON SCANDENS)

The heartleaf philodendron is a top choice for newbies because it's low-maintenance and hard to kill, but it is lovely no matter what your skill level is. As its name suggests, it has heart-shaped leaves that are a glossy dark green. It's a climbing plant and will thrive in a hanging basket or fastened to a pole or trellis.

While the heartleaf philodendron will thrive in moderately bright light, it will also make itself at home in a spot that doesn't get any direct sunlight. It's hardy, so you can test out that lower-light spot, and if it starts to look sad just move it somewhere else.

How to Make Friends
with Heartleaf Philodendrons

» SOIL

Use an all-purpose potting mix.

» LIGHT

Bright indirect or filtered light is ideal, such as from an eastern or western window, but the lower-light conditions that a northern window provides should be fine too.

» WATER AND HUMIDITY

Allow soil to dry out slightly before watering, to a depth of 1–2 inches (2.5–5 cm). Yellowing leaves may indicate overwatering. Average humidity conditions are fine.

» TEMPERATURE

Room temperatures of 60–80°F (16–27°C) are suitable for this plant.

» SIZE

A heartleaf philodendron can get quite large, with stems 4–8 feet (1.2–2.4 m) long, making it eventually suitable as a floor plant, but you can cut it back to keep it smaller and more manageable in a hanging basket or regular pot.

» BUGS AND DISEASE

Watch out for root rot (page 44), aphids (page 39), and scale (page 41).

If you use a pole or trellis, you'll need to physically attach the plant with supports such as plant ties or twist ties. It will develop aerial roots that will eventually latch on, allowing you to remove the supports.

As you would do with some other plants with broad, flat leaves, occasionally dust the leaves with a damp cloth. Alternatively, if the plant is small, give it a shower with room-temperature water.

A heartleaf philodendron can easily be propagated with stem-tip cuttings (page 51).

INCH PLANT
(TRADESCANTIA ZEBRINA)

Native to Mexico, the inch plant creeps on the ground when it's in an outdoor environment, sometimes even becoming invasive in certain climates. Indoors, it is a favorite hanging basket plant, though you can certainly grow it in a regular pot. Its richly toned leaves have silvery-green and purple stripes.

When you get your plant, it may be small and bushy, but it will eventually grow long trailing stems (2–3 feet/60–90 cm long). To keep it looking spiffy and full, prune it. Overall, it's a fuss-free choice as a houseplant and super-easy to propagate with stem-tip cuttings (page 51). If you're looking for some other options, check out the popular pink variegated cultivar called *T. fluminensis* 'Tri-color'.

How to Make Friends with Inch Plants

≫ SOIL
Use an all-purpose potting mix.

≫ LIGHT
Grow your inch plant in bright light, such as that from an east or west window. It will also do well in a slightly more moderate light, such as filtered or indirect light.

≫ WATER AND HUMIDITY
Keep the soil consistently moist, watering when the top inch of the soil feels dry. Water less often in winter. Average levels of humidity are fine.

≫ TEMPERATURE
Room temperatures of 60–80°F (16–27°C) are suitable for this plant.

≫ SIZE
Although this small- to medium-sized plant doesn't get very tall (around 6 inches/15 cm or so), it does have long, trailing stems.

≫ BUGS AND DISEASE
Watch out for spider mites (page 42), root rot (page 44), and aphids (page 39).

≫ OTHER TIPS
The inch plant can easily be propagated with stem-tip cuttings (page 51). You can also divide your plant and pot up the divisions separately.

JADE PLANT (CRASSULA OVATA)

The jade plant is a well-loved—some even say lucky—houseplant. This undemanding, beautiful succulent has fleshy jade-green leaves, and as it matures, it looks like a miniature tree. Like many other succulents, the jade plant doesn't require a lot of water or humidity and can be propagated from a single leaf cutting (page 50). (It's often called a *friendship plant* because you can share the baby plants with others.) Jade plants can live for quite a long time, and if you're lucky enough to keep yours to a ripe age of ten years or so, it can flower in winter. If you think that the jade plant looks a bit bonsai-like, you have a good eye. It is often used as bonsai, because it's easy for beginners to prune into shape.

How to Make Friends with Jade Plants

>> **SOIL**

Use a cactus and succulent mix.

>> **LIGHT**

A jade plant will be happy in bright light, such as that from a western or southern window. In full sun, the leaves may turn slightly reddish. If the plant is starting to look leggy (long and slender) or leans toward the light source, it needs more light.

>> **WATER AND HUMIDITY**

Let the soil dry out a little bit before watering, to a depth of 1–2 inches (2.5–5 cm). If the leaves turn yellow, you're probably overwatering. Water less frequently in winter. There are no particular humidity requirements.

>> **TEMPERATURE**

Room temperatures of 60–80°F (16–27°C) are suitable for this plant.

>> **SIZE**

Jade plants can vary in size depending on age. A new plant may fit on a windowsill, while a mature plant may grow to 2–3 feet (60–90 cm) tall.

>> **BUGS AND DISEASE**

Watch out for root rot (page 44) and mealybugs (page 40).

>> **OTHER TIPS**

When you first buy a jade plant, there may be several plants in the pot. These should be divided up and placed into individual pots in the spring. Wipe down the leaves occasionally with a damp cloth to remove dust.

LADY SLIPPER ORCHID (PAPHIOPEDILUM)

I can already hear you say it. *I cannot possibly grow this plant. It's too gorgeous. It looks so difficult.* Looks can be deceiving. Expert after expert will tell you that this is one of the easiest orchids, even for beginners. You can do it. Really.

While you might be quite familiar with the moth orchid (page 121)—which you can probably buy by the cartload at your local supermarket—you may be less familiar with the lady slipper. The name refers to the pouch-like appearance of the flower. Unlike many other orchids, *Paphiopedilums* are not epiphytes (air plants), but terrestrial plants. They don't require a lot of light, making them a good choice for almost anyone. One of the best things about many orchids, including this one, is that the flowers last forever, or at least for several months.

How to Make Friends with Lady Slipper Orchids

» SOIL

Use a semi-terrestrial orchid mix (if available) or a combination of sphagnum moss and orchid bark.

» LIGHT

Indirect light is best for these orchids. Near an eastern window is a good choice, as is a northern window as long as you can get the plant quite close to the window. A south- or west-facing arrangement will require filtered light and/or moving the plants back from the window, as these orchids are prone to sunburn.

» WATER AND HUMIDITY

Lady slipper orchids require more water than some other orchids. Water when the top of the potting mix feels dry to the touch, usually every few days. Ideally, you should use filtered or distilled water for these lovelies. These tropical plants like humidity, so use a pebble tray (page 18) to increase humidity around the plant.

» TEMPERATURE

Room temperatures of 60–85°F (16–30°C) are ideal for this plant, but it can tolerate temperatures outside this range.

» SIZE

This plant is relatively small, with leaves about 6 inches (15 cm) long. A flower spike will usually be more than 12 inches (30 cm) tall.

» BUGS AND DISEASE

Watch out for sunburn (page 47) and mealybugs (page 40).

» OTHER TIPS

Want your orchid to rebloom? Make sure there's a 15°F (8°C) temperature difference between night and day.

LEMON-LIME DRACAENA (DRACAENA DEREMENSIS 'LEMON LIME')

I was drawn to the brightly striped leaves of lemon-lime dracaena—also known as corn plant—even before I knew what this plant was called. I plopped my mystery plant in front of a window that didn't have particularly good light and left it there. I intended to do something with it eventually—like perhaps learn what it was and how to care for it—but I never quite got around to it. It didn't seem to mind. It didn't grow much, but the neglect didn't seem to hurt it much either.

With better treatment and better light, it can grow much taller (several feet!), but the small size and unfussiness of this cheerful plant can be appealing as well. It looks good in many different kinds of decor, adding a welcome pop of color, and while I hope you'll treat yours a bit better than I did mine, know that it won't suffer too much if you don't.

How to Make Friends
with Lemon-Lime Dracaenas

≫ **SOIL**

Use an all-purpose potting mix.

≫ **LIGHT**

Lemon-lime dracaena can survive a wide range of light conditions but will do best with moderate to bright light, such as an east- or west-facing window. It will grow, but not thrive, in a north-facing window.

≫ **WATER AND HUMIDITY**

Let the soil dry out slightly before watering. In dry conditions, add a pebble tray (page 18) to provide additional humidity. Dry leaf tips can mean underwatering or insufficient humidity.

≫ **TEMPERATURE**

Room temperatures of 60–80°F (16–27°C) are ideal, but the plant can tolerate conditions outside this range.

≫ **SIZE**

A young plant may be around 12 inches (30 cm) tall and suitable for a table, but it can become a larger floor plant (over 5 feet/1.5 m tall) with age. Larger plants are also easy to find for purchase.

≫ **BUGS AND DISEASE**

Watch out for mealybugs (page 40), root rot (page 44), and scale (page 41).

≫ **OTHER TIPS**

Like many plants with long, wide leaves, this one gathers dust, so give it an occasional shower in lieu of a watering, or dust the leaves with a damp cloth.

LUCKY BAMBOO
(DRACAENA SANDERIANA)

You're almost certainly familiar with lucky bamboo. It's common in restaurants, offices, and stores. Because it's thought to bring happiness and, well, luck, it's often given as a gift. You would be forgiven for not even considering it a houseplant because it's usually grown in water, though it can grow in soil. You would also be forgiven for thinking it's bamboo. It's actually a dracaena, a genus that also includes the lemon-lime dracaena (page 112). The bamboo-like stems are often trained into twisted or braided shapes. If you purchase or receive a plant growing in water, you can leave it there; just be sure to keep the roots covered. The only trick to these plants is to use distilled, filtered, or rainwater. They don't like the chlorine in most tap water. Otherwise, it's one of the easiest houseplants you can find: hard to kill, doesn't need much light, and doesn't even require soil. May it bring you good luck!

How to Make Friends with Lucky Bamboo

>> SOIL

Lucky bamboo usually grows in water, but you can also plant it in all-purpose potting mix.

>> LIGHT

It will do best in low to moderately bright light, such as that from a northern or eastern window, but feel free to experiment with even lower-light situations. Just keep it out of direct sunlight.

>> WATER AND HUMIDITY

If you grow the plant in water, change the water every week or so, using distilled, filtered, or rainwater. Add a drop of fertilizer every few months. If grown in soil, keep the soil moderately moist but never soggy, and boost humidity in dry rooms with a pebble tray (page 18).

>> TEMPERATURE

Room temperatures of 60–75°F (16–24°C) are suitable for this plant.

>> SIZE

Lucky bamboo can come in various sizes, from a few inches in height to 3 feet (90 cm).

>> BUGS AND DISEASE

Watch out for mealybugs (page 40).

>> OTHER TIPS

If the water smells, change it more often! If you grow lucky bamboo in water, you can add pebbles or marbles for decoration.

MEYER LEMON TREE (CITRUS × MEYERI)

There is nothing quite like growing your own citrus indoors. You don't need to live somewhere semi-tropical or tropical to do so, and dwarf plants, which are now quite widely available at garden centers and nurseries, stay a manageable size in the home.

Considered to be a cross between a lemon and a mandarin, the Meyer lemon has less acidity than a regular supermarket lemon and a slightly floral flavor. It's a highly coveted (and often expensive) ingredient in desserts and other dishes, so why not grow your own? This plant wants a lot of light, so place it in front of a southern-facing window if possible. If you have an outdoor space, such as a balcony, put it outdoors in the summer. A Meyer lemon tree produces lots of flowers that smell glorious and will start to bear fruit when it's about two or three years old.

How to Make Friends with Meyer Lemon Trees

>> SOIL

Use an all-purpose potting mix that drains very well, or a citrus potting mix if you can find one.

» LIGHT

Meyer lemon trees love the sun, so put them in a place with full sun, such as a south-facing window. If you have outdoor space, put them outdoors in summer.

» WATER AND HUMIDITY

Water when the top inch (2.5 cm) of the soil feels dry, and keep the watering schedule consistent, not letting it dry out too much. Water less frequently in fall and winter. For additional winter humidity, you can set your plant on a pebble tray (page 18).

» TEMPERATURE

Room temperatures above 60°F (16°C) are suitable for this plant.

» SIZE

A Meyer lemon tree can be 1–3 feet (30–90 cm) tall.

» BUGS AND DISEASE

Meyer lemon is prone to scale (page 41). It can also get mealybugs (page 40).

» OTHER TIPS

Air circulation will help pollination. I sometimes run my fingers over the blossoms as well for good measure.

You will need to fertilize your tree during the growing season, from early spring through early summer. Look for an organic citrus fertilizer that you can sprinkle on top of the soil.

MONSTERA
(MONSTERA DELICIOSA)

This "delicious monster," also known as the split-leaf philodendron and Swiss cheese plant, is the stuff of stylish design blogs and enviable social media feeds. In Latin, *monstera* means "abnormal," and there is something unusual—and unusually appealing—about the plant's look and its show-stopping split leaves. It does produce an edible fruit, but it rarely does so indoors. The rest of the plant is toxic, so don't go gnawing on it—and keep pets away too.

It's easy to fall in love with monstera, and this floor plant is easy to grow, but be careful: the green monster can get to be quite large. Don't be upset if your monstera doesn't display its distinctive split-leaf pattern right away. The leaves of younger plants are solid; as the plant ages, they'll start to get holes, or *fenestrations*, if you want to get fancy.

How to Make Friends with Monstera

>> SOIL

Use an all-purpose potting mix.

>> LIGHT

This plant likes bright indirect or filtered light, such as that from an eastern window, but you can also grow it in lower light conditions, such as that from a north-facing window. In its native habitat, this plant would be growing under tall trees, with light filtering down through the leaves. For this reason, it initially grows *away* from the light, a phenomenon called *negative phototropism*, so that it can find the shade of a tree. Avoid direct sun, and rotate the plant for even growth.

>> WATER AND HUMIDITY

Water when the first inch (2.5 cm) or so of the soil feels dry to the touch. In fall and winter, water less often. As a tropical plant, monstera likes humid air. To boost humidity, use a pebble tray (page 18).

>> TEMPERATURE

Room temperatures of 60–85°F (16–30°C) are suitable for this plant.

>> SIZE

A small monstera might be 1½ feet (45 cm) tall, but it can grow to 6–8 feet (1.8 to 2.4 m). This is definitely a floor plant.

» BUGS AND DISEASE

Monstera is not prone to serious insect or disease problems, but look out for aphids (page 39), mealybugs (page 40), scale (page 41), or root rot (page 44).

» OTHER TIPS

This plant has *aerial roots*, or roots that grow aboveground. You can put the lower roots into the soil to help nourish the plant, and then train the higher-up roots on a pole or trellis.

If you don't have the space for a whole monstera plant, you can decorate with cuttings. A few leaves in a large vase will last a month or two, adding temporary tropical flair to your home.

MOTH ORCHID *(PHALAENOPSIS)*

Orchids are one of the biggest flowering-plant families in the world, with more than 25,000 individual species across more than 700 genera. Of these, the most popular variety to grow as a houseplant is undeniably the moth orchid. There's something gracefully imperfect about a moth orchid: the way its leaves tilt, the twists of its greedy reaching tendrils, and how the flowers seem to float high above the plant. It's no wonder that orchids have inspired plant lust (and theft!) throughout history. Even Charles Darwin was entranced by orchids, which he discovered had perfectly evolved alongside their pollinators. He considered this conclusive evidence that supported the theory of evolution.

Orchids can be quite easy to grow, despite their reputation to the contrary. Moth orchids are epiphytes (air plants), so they should be

potted in bark, coco chips, or sphagnum moss instead of soil. They often flower in winter, and the flowers can last for months.

How to Make Friends with Moth Orchids

>> SOIL

Use an orchid potting mix, coco chips (shredded coconut husks), or sphagnum moss.

>> LIGHT

The moth orchid likes a moderate level of light. Place it near an east-facing window, or provide filtered light near a west-facing window. If you have a bathroom that provides enough light, an orchid will be happy with the humidity levels. Avoid direct sun.

>> WATER AND HUMIDITY

Allow the potting mix to dry out before watering. In most cases, the plants can be watered once a week. Take the plant to a sink and water thoroughly under tepid running water. Let drain completely and return the plant to its location. Alternatively, you can also soak the plant in room-temperature water for 20 to 30 minutes. If humidity is low where you live, put the plant on a pebble tray (page 18).

>> TEMPERATURE

Room temperatures of 60–85°F (16–30°C) are suitable for this plant.

Moth orchids are relatively small plants, suitable for a windowsill or tabletop.

≫ BUGS AND DISEASE

Moth orchids are not particularly prone to pests or disease, but watch for scale (page 41).

≫ OTHER TIPS

Orchid roots will grow outside the pot; just let them do their thing and don't cut them off unless they seem shriveled and hollow. A healthy root will look bright green with a silvery sheath.

For an orchid to rebloom, it needs to be exposed to slightly cooler temperatures. You can place your orchid outside at night for several weeks in late summer or early fall, as long as the temperatures don't dip below 55°F (13°C). You can also leave it on a cool windowsill in winter. It can take months between the appearance of a flower spike and first flower opening, so be patient.

How can you tell the difference between a flower spike and an aerial root? An emerging flower spike will look like a little mitten.

Occasionally, a moth orchid will produce a new plant, called a *keiki*, on an old flower stalk. Once this plant gets some leaves and roots, you can cut it off and pot it up as a new plant.

NERVE PLANT (FITTONIA)

If you prefer to dote on your plant friends, the nerve plant may be just the one for you, especially if you live in a warm, humid location. This South American rain forest native loves high humidity and consistently moist soil. If it doesn't get enough water, it may have a fainting spell and wilt—putting the "fit" in "fittonia." And if it gets too much water, it may get root rot. Like Goldilocks, you want to get this one *just right*. Despite these challenges, fittonias easily captivate with their delicately veined leaves lined in white, pink, or red. If providing them the environment they need is difficult in your home, don't worry: they thrive in terrariums and look great paired with polka dot plants (page 136), which have similar care requirements.

How to Make Friends with Nerve Plants

>> SOIL

Use all-purpose potting mix.

>> LIGHT

A nerve plant likes moderate to bright indirect or filtered light from an eastern or western window. Keep out of direct sun.

>> WATER AND HUMIDITY

Keep the soil moderately moist, but not soggy. Consider seeking out a miniature version to plant in a terrarium to manage the humidity and temperature requirements more easily. Use a pebble tray (page 18) or humidifier to increase humidity.

>> TEMPERATURE

Room temperatures of 65–80°F (18–27°C) are suitable for this plant. Temperatures below 55°F (13°C) should be avoided.

>> SIZE

Nerve plants are quite small, between 3 and 6 inches (7.5–15 cm) tall.

>> BUGS AND DISEASE

Watch out for aphids (page 39) and root rot (page 44).

>> OTHER TIPS

Nerve plants can be propagated with stem-tip cuttings (page 51).

NORFOLK ISLAND PINE (ARAUCARIA HETEROPHYLLA)

If you love the idea of having a Christmas tree year-round or bringing the forest indoors, a Norfolk Island pine is the plant for you. Originally from Australia's Norfolk Island, these plants—which are not true pine trees—are often sold seasonally as miniature Christmas trees. There's a lot more to recommend them than holiday cheer, however. A Norfolk Island pine can add a wonderfully rustic feel to your home.

As a houseplant, this tree is usually about 2 feet (70 cm) tall when you purchase it, and it will slowly grow to 6 feet (1.8 m) after many years. Although you can probably pick one up at any supermarket, big-box, or warehouse store in December, you're often better off going to a reputable garden center, because some of the seasonal trees are sprayed with green paint or even glitter.

How to Make Friends
with Norfolk Island Pines

» SOIL

Use an all-purpose potting mix, ideally mixing in some sand to improve drainage.

» LIGHT

A Norfolk Island pine likes bright light with some direct sun, such as from an eastern, western, or southern window. If the needles start turning yellow, it is likely getting too much light. Rotate the plant a quarter-turn weekly to promote even growth.

» WATER AND HUMIDITY

Keep the soil moderately moist, but not soggy. Use a pebble tray (page 18) to increase humidity, and also mist occasionally. If the room is dry, a humidifier will help tremendously.

» TEMPERATURE

Room temperatures of 55–75°F (13–24°C) are suitable for this plant.

» SIZE

Norfolk Island pines can reach 6 feet (1.8 m) tall.

» BUGS AND DISEASE

Watch out for mealybugs (page 40), root rot (page 44), spider mites (page 42), and scale (page 41).

OXALIS TRIANGULARIS

With deep-purple leaves that float like butterflies on long antenna stems, *Oxalis triangularis* is gracefully delicate in appearance, but tougher than it looks. Like the prayer plant (page 142), it responds to light in an entertaining way: in the morning, its three leaflets open like an umbrella, and then they shut again each evening.

You may hear this plant called "purple shamrock," "shamrock plant," or "false shamrock," but it's from Brazil, not the Emerald Isle. Because it grows from underground bulbs called corms, the foliage occasionally dies back and the plant goes dormant. When this happens, give the plant a rest; don't water or do anything to it. It will start to send up new shoots in two to four weeks (sometimes longer) and then you can resume care again.

How to Make Friends
with *Oxalis Triangularis*

>> **SOIL**

Use an all-purpose potting mix.

>> **LIGHT**

This plant likes medium to bright indirect light. Place it near
an east-facing window, or give it filtered light from a brighter
window.

>> **WATER AND HUMIDITY**

Water when the top inch (2.5 cm) of the soil feels dry. Once the
plant goes dormant, don't water until you see new growth. Average
room humidity is fine for this plant.

>> **TEMPERATURE**

Room temperatures of 60–75°F (16–24°C) are suitable for this
plant.

>> **SIZE**

This plant is rather small, around 6 inches (15 cm) tall.

>> **BUGS AND DISEASE**

Watch out for spider mites (page 42).

>> **OTHER TIPS**

If the foliage dies due to lack of watering, simply cut off the dead
leaves, let the plant fall dormant, and wait for it to start up again.

PANDA PLANT
(KALANCHOE TOMENTOSA)

This furry friend, also called the *teddy bear plant*, is soft, soft, soft, and super simple to care for. It's a succulent, so you don't need to water it that frequently, and it's perfectly happy in a dry room. The only thing it asks for is some time in the sun. (Don't we all?) It can even tolerate full sun if you acclimate it gradually. When you buy the plant, it may be quite small and perfect for your windowsill, although it can eventually grow taller. If you want to create your own succulent dish garden, this plant is a nice choice, because its fuzzy leaves will provide variety and texture.

How to Make Friends with Panda Plants

» SOIL

Use a cactus and succulent potting mix.

» LIGHT

A panda plant likes bright light, such as that from a western or southern window. If you want to grow the plant in full sun, gradually expose the plant to sunnier conditions and watch for sunburn.

» WATER AND HUMIDITY

Allow the top inch (2.5 cm) or so of the soil to dry out before watering. This plant will tolerate a missed watering much more than it handles overwatering, which could cause root rot. Avoid getting water on the fuzzy leaves. It doesn't mind dry air.

» TEMPERATURE

Room temperatures of 60–80°F (16–27°C) are suitable for this plant.

» SIZE

Size can range from a few inches to 2 feet (60 cm) tall.

» BUGS AND DISEASE

Watch out for mealybugs (page 40), root rot (page 44), and sunburn (page 47).

» OTHER TIPS

Like other succulents, the panda plant is easy to propagate with leaf cuttings (page 50).

PARLOR PALM (CHAMAEDOREA ELEGANS, SYN. NEANTHE BELLA)

No book on popular houseplants would be complete without at least one true palm. And the classic parlor palm, with its feathery leaves, is one of the easiest to grow. Its popularity dates back to Victorian times, when palms and other sturdy plants decorated poorly lit parlors. A palm tree can be a great indoor floor plant for almost any design situation, adding height, texture, and tropical lushness to a room. Because

it can tolerate relatively low light, it won't protest if you stick it in a corner, provided that the corner does get some bright indirect light. A parlor palm also grows slowly and tops out at a manageable size, so it won't take over your room.

How to Make Friends with Parlor Palms

>> SOIL

Use an all-purpose potting mix or a cactus and succulent mix.

>> LIGHT

Parlor palms can tolerate low light, such as from a north-facing window, but would prefer brighter, filtered light. Avoid direct sun.

>> WATER AND HUMIDITY

Water when the top inch (2.5 cm) of the soil feels dry. Place the plant on a pebble tray (page 18) to increase humidity and help prevent spider mites.

>> TEMPERATURE

Room temperatures of 60–80°F (16–27°C) are suitable for this plant.

>> SIZE

The parlor palm can reach 3–4 feet tall (90–120 cm), making it a good choice as a floor plant.

>> BUGS AND DISEASE

Watch out for spider mites (page 42).

>> OTHER TIPS

If possible, give your palm an occasional shower (perhaps a few times a year, or whenever it gets very dusty), with tepid water to rinse the leaves of dust and help prevent spider mites. It will thank you for the extra burst of humidity!

If you're looking for a taller palm with even more tropical feel, try a Kentia palm (*Howea forsteriana*), which can grow to be 6–7 feet (1.8–2.1 m) tall; it has similar care requirements.

PEACE LILY *(SPATHIPHYLLUM)*

Even before a NASA study named the peace lily one of the top air-purifying plants, it was a perennial favorite in the home. It's low-maintenance and doesn't need a ton of light to grow. It also has attractive glossy leaves and as long as the plant is receiving sufficient light, it will periodically produce pretty white blossoms that last about a month. While a peace lily prefers a consistent watering schedule, it will bounce back if underwatered. When the plant really needs water, it will droop. It will perk up again once it's properly hydrated, but to avoid stressing the plant, try not to let this happen too often.

How to Make Friends with Peace Lilies

» **SOIL**

Use an all-purpose potting mix.

» **LIGHT**

A peace lily can grow in low light, such as that from a northern window, but will be happier in bright filtered light. Keep it out of direct sun.

» **WATER AND HUMIDITY**

Maintain a consistent watering schedule, watering when the top inch (2.5 cm) of the soil feels dry. If your water is highly chlorinated, use filtered water. Average room humidity is fine for this plant.

» **TEMPERATURE**

Room temperatures of 60–85°F (16–30°C) are suitable for this plant.

» **SIZE**

Small peace lilies are around 1 foot (30 cm) tall. Larger varieties can reach 4 feet (1.2 m) tall and can be grown as a floor plant.

» **BUGS AND DISEASE**

Watch out for mealybugs (page 40) and scale (page 41).

» **OTHER TIPS**

If you have one of the larger varieties of the peace lily, dust the leaves occasionally with a damp cloth.

POLKA DOT PLANT
(HYPOESTES PHYLLOSTACHYA)

Like the nerve plant (page 124), the polka dot plant has super-pretty, delicately patterned foliage and is a perfect choice for a terrarium because it loves humidity. The most popular polka dot plants have pink markings, though you can also find ones with white, light purple, or red spots. As the stems of your polka dot plant grow longer, prune some of the new growth (this is called *pinching off* because you can pinch the stem and leaves off with your thumb and forefinger) to produce a bushier, more attractive plant.

One important thing to know: this plant doesn't live a long life, two years max. Once it flowers, it's usually a goner, but sometimes it dies without even flowering. Don't be sad—just get yourself a new plant.

How to Make Friends with Polka Dot Plants

» SOIL

Use an all-purpose potting mix.

» LIGHT

Polka dot plants like bright light but can adapt to moderate light. Filtered light through a sheer curtain on a south-facing window will work well. East- and west-facing windows are good too, though you might move the plant back slightly from a west window. If the leaves start to curl or turn brown, it's most likely getting too much light.

» WATER AND HUMIDITY

Don't let the soil dry out entirely; keep it consistently moist but never soggy. Water when the top half inch to 1 inch (1–2.5 cm) feels dry. A polka dot plant likes humidity and lots of it. Place it on a pebble tray (page 18) and/or group with other plants to increase the humidity levels around it. You can also mist your plant frequently. Alternatively, grow it in a terrarium.

» TEMPERATURE

Room temperatures of 60–80°F (15–27°C) are suitable for this plant.

» SIZE

Polka dot plants can reach a height of 12–18 inches (30–45 cm).

» BUGS AND DISEASE

Watch out for powdery mildew (page 45) and whiteflies (page 42).

» OTHER TIPS

Try picking up a nerve plant and a polka dot plant in contrasting or complementary colors and growing them together in a terrarium. Add a frilly fern for balance and texture.

You can grow polka dot plants from seed!

PONYTAIL PALM
(BEAUCARNEA RECURVATA)

The ponytail palm is not actually a palm, but its mop of long leaves does look like a ponytail. (For a true palm, see the parlor palm on page 132.) The most striking feature of this plant is its slim trunk, which bulges out at the bottom; this base is called a *caudex* and helps the plant retain water. (If you've been following along so far, you'll pick up the hint that this plant's appearance offers: "Don't overwater me!" it's tell-

ing you.) Because of the caudex, the ponytail palm also goes by the name *elephant foot*. Outdoors, it grows in semi-desert climates, so to re-create that environment at home, give it lots of sun and a quick-draining soil, avoid overwatering, and put down your mister. The low water requirements make it an excellent choice for frequent travelers and those who tend to forget about their plants.

How to Make Friends with Ponytail Palms

>> SOIL

Use a cactus and succulent potting mix.

>> LIGHT

A ponytail palm needs a lot of light, ideally from a south- or west-facing window. If the tips of the leaves turn brown, this can be a sign that the plant is getting too much sun. After moving the plant to a shadier location, you can clip off the brown sections to keep your ponytail palm well groomed.

>> WATER AND HUMIDITY

A ponytail palm stores a lot of water in its base, so wait until the soil is mostly dry (to a depth of at least 2 inches/5 cm) before watering. Water less frequently in winter. No supplemental humidity is needed for this desert plant.

>> TEMPERATURE

Room temperatures of 60–80°F (16–27°C) are suitable for this plant.

>> SIZE

Ponytail palms range in size from smaller plants suitable for a tabletop to floor plants of 6 feet (1.8 m) and up.

>> BUGS AND DISEASE

Watch out for root rot (page 44) and spider mites (page 42).

>> OTHER TIPS

The ponytail palm likes to be in a fairly compact and shallow pot for its size.

POTHOS *(EPIPREMNUM AUREUM)*

Pothos is known by many names, including golden pothos, money plant, and devil's ivy. For some people it's a bit yawn-worthy because it's so common in commercial or office settings, but it's quite beautiful and one of the most laid-back plant friends you'll ever make because it's so easy to care for. It doesn't need a lot of light and can survive occasional neglect. Pothos also works for you: it's one of NASA's top air-purifying plants.

The most popular variety, *Epipremnum aureum*, has variegated green and yellow heart-shaped leaves. A trailing plant like pothos can add balance and variety to any houseplant collection. Put it on a high shelf and let the long vines fall where they may, train it on a trellis or other support, or even cut it back to make it bushy. To create a lot of indoor drama, pair it with one big floor plant. Add a few smaller favorites, and you've got yourself a perfect minimalist collection.

How to Make Friends with Pothos

» SOIL

Use an all-purpose potting mix.

» LIGHT

Pothos prefers medium to bright filtered light, such as that from an eastern or western window. In lower-light conditions, such as northern exposure, it may grow more slowly and the leaves may become greener and less variegated.

» WATER AND HUMIDITY

Water when the top inch (2.5 cm) of the soil feels dry. Your goal is to keep the soil both from drying out completely and from getting soggy. If the leaves turn yellow, you're likely overwatering. Average room humidity levels will keep pothos happy.

» TEMPERATURE

Room temperatures of 60–80°F (16–27°C) are suitable for this plant.

» SIZE

Pothos is a climbing or trailing plant. The vines can grow very long (more than 10 feet/3 m).

» BUGS AND DISEASE

Watch out for root rot (page 44) and mealybugs (page 40).

» OTHER TIPS

New leaves are green when they first emerge, but they develop variegation with time.

If the vines become too long, don't be afraid to trim them back to keep the plant the size you want.

If the itch for propagation strikes, place a stem-tip cutting (page 51) of pothos in a bud vase and wait for it to grow roots.

PRAYER PLANT
(MARANTA LEUCONEURA)

Along with *Oxalis triangularis* (page 128), the prayer plant falls into that unscientific category that I like to call "plants that do stuff." Like oxalis, it moves in response to light, or lack thereof, beatifically opening its leaves at dawn and closing up shop at night. It's often said that the closed leaves look like a person's hands in prayer, hence the common name "prayer plant." The other attraction of this easy-to-find plant is its uniquely patterned leaves. Different varieties have different patterns, but they're all striking and unusual, complementing almost any kind of decor. You might think that such a peculiar plant might be hard to care for, but it's surprisingly low-maintenance. It doesn't need lots of light; in fact, it will even grow in a north-facing window. If you can give it some extra humidity, through misting or a pebble tray, it will be happier, but drier conditions are not a complete deal-breaker.

How to Make Friends with Prayer Plants

>> SOIL

Use an all-purpose potting mix.

>> LIGHT

A prayer plant prefers a medium level of indirect or filtered light. You can place it in a north-facing window or near an east-facing window with a sheer curtain. Avoid direct sun.

>> WATER AND HUMIDITY

Keep the soil moderately moist but not soggy, watering when the top half inch (1 cm) of the soil feels dry. Water the plant less frequently in winter. If you're putting your plant in a dry room, use a pebble tray (page 18) or place it close to other plants. If your bathroom has sufficient light, you can also place your plant there for a more humid environment.

>> TEMPERATURE

Room temperatures of 60–80°F (16–27°C) are suitable for this plant.

>> SIZE

A prayer plant is relatively small, growing to about a foot (30 cm) or so in height.

>> BUGS AND DISEASE

Watch out for spider mites (page 42).

>> OTHER TIPS

Your prayer plant will likely go through a rest period in winter, when its growth will slow.

REX BEGONIA *(BEGONIA REX)*

Rex begonias are all about the leaves: the foliage is a riot of purples, reds, greens, and other colors. It's no wonder they're often called "painted-leaf" or "fancy-leafed" begonias. While you don't need a lot of light to grow these plants, you'll want to keep the humidity levels high, ideally 50 percent or more. If you can't easily create those conditions in your home, you can put a rex begonia in a terrarium like those other steam fiends, the nerve plant (page 124) and polka dot plant (page 136). Note that these beauties can go dormant in the winter. If this happens, they'll look rather sad and lose most of their leaves, but go with the flow: just cut back the dead leaves and let the plant rest, giving it very little water until spring. (Now, if it goes "dormant" at another time, you might simply have a dead plant!)

How to Make Friends with Rex Begonias

≫ **SOIL**

Use an all-purpose potting mix.

≫ **LIGHT**

Place your rex begonia where it will receive bright indirect or filtered light, such as an eastern or northern window. Avoid direct sun.

≫ **WATER AND HUMIDITY**

Consistent watering is crucial for rex begonia. Let the soil dry out slightly on top before giving the plant water. After watering, the soil should be moderately moist but not soggy. Reduce the frequency of watering in the winter, particularly if the plant is dormant. Rex begonias are humidity hogs. Place plants on a pebble tray (page 18) to increase humidity or grow in a terrarium.

≫ **TEMPERATURE**

Room temperatures of 60–75°F (16–24°C) are ideal for this plant.

≫ **SIZE**

Rex begonias are about 8–12 inches (20–30 cm) tall.

≫ **BUGS AND DISEASE**

Watch out for mealybugs (page 40), powdery mildew (page 45), and gray mold (page 43).

≫ **OTHER TIPS**

These plants prefer shallow pots. Keep that in mind when it's time to repot your rex.

Avoid misting your rex begonia or giving it a shower. It doesn't like wet leaves; the water could cause spots to form and lead to disease.

RUBBER PLANT *(FICUS ELASTICA)*

The rubber plant, or rubber tree, is a ficus just like the fiddle-leaf fig (page 96)—and it's even easier to care for. It doesn't need a lot of light, is not particularly susceptible to pests, and won't drop its leaves at the slightest change in conditions, like some other ficuses commonly grown as houseplants. It's also a good air-purifying plant. The large leaves are satisfyingly thick, leathery, and shiny, and something about the plant just exudes a sense of stability. The 'Burgundy' variety has a blackish, reddish-green leaf color that almost defies description. 'Tineke' has gorgeous variegated leaves of splotchy light and dark green edged with yellow, while 'Ruby' has slightly pinkish edges. Research several varieties and pick your favorite.

How to Make Friends with Rubber Plants

≫ SOIL

Use an all-purpose potting mix.

>> LIGHT

A rubber plant likes bright indirect light, such as that from an eastern or western window.

>> WATER AND HUMIDITY

Water when the top inch (2.5 cm) of the soil feels dry. The soil should be evenly moist, but the plant is susceptible to root rot with overwatering. Water less frequently in winter. Consider a pebble tray (page 18) if the room is dry.

>> TEMPERATURE

Room temperatures of 60–80°F (16–27°C) are suitable for this plant.

>> SIZE

In the wild, this tree can grow between 50 and 100 feet (15 and 30 m) tall, but as a houseplant it's unlikely to be more than 10 feet (3 m) tall, and often much shorter, after many years of growth.

>> BUGS AND DISEASE

Watch out for root rot (page 44) and scale (page 41).

>> OTHER TIPS

Keep the leaves clean and shiny; dust them periodically with a damp cloth.

Wear gloves when handling a rubber tree. When you nick or prune it, it will exude a milky sap, which is a skin irritant. You can stop the bleeding by spraying the spot with water or placing a damp paper towel on it.

SNAKE PLANT
(SANSEVIERIA TRIFASCIATA)

Few plants say "set it and forget it" like the snake plant. Subsisting on little more than hardship and grit, the snake plant doesn't even require light or water. Okay, not true, but it is one of the easiest houseplants to grow successfully, especially for neglect- ful owners. This succulent—yes, it's a suc- culent!—is a plant that can survive drought conditions, and its long, sturdy, spiky leaves are striking. One of the most common vari- eties, 'Laurentii', has variegated green leaves with yellow edges. Less common but still easy to find is the stunning 'Moonglow' (sometimes called 'Moonshine'); its pale silvery-green leaves edged in dark green have a washed-out watercolor feel to them. Consider spacing several across the top of a media console or buffet in identical con- tainers, choose a few different cultivars and cluster them together, or pick up a set of matching planters in different heights to create visual interest. This plant is also a great air purifier, another reason to have a few of them around.

How to Make Friends with Snake Plants

>> SOIL

Use potting mix that drains well, such as a cactus and succulent mix.

>> LIGHT

Bright light is ideal for the plant to thrive, but a snake plant can adapt to a range of other light conditions, including low light. Constant direct sun could burn the leaves though.

>> WATER AND HUMIDITY

Let the top inch or two (2.5 or 5 cm) of the soil dry out before watering. Water less frequently in winter. This plant has no special humidity needs.

>> TEMPERATURE

Room temperatures of 60–80°F (16–27°C) are suitable for this plant.

>> SIZE

Snake plants are generally between 1 and 3 feet (30 and 90 cm) tall.

>> BUGS AND DISEASE

Problems with a snake plant are unlikely, but it could get mealybugs (page 40), spider mites (page 42), or root rot (page 44).

>> OTHER TIPS

Because these plants are taller than they are wide, use a wide, sturdy pot to keep them from toppling.

Snake plants are easy to propagate from leaf cuttings (page 50). Cut a leaf into segments a few inches long. Make sure you keep track of which is the top and bottom of the segment—the plant knows which end is up, and if you plant the segments upside down, they won't grow. Place the segments bottom down into a potting medium.

SPIDER PLANT
(CHLOROPHYTUM COMOSUM)

The common spider plant—some detractors might say *too* common—is a great houseplant for beginners and is also terrific at purifying air. That alone makes it worth a second look, even if the mere thought of another spider plant in the world makes you want to turn the page in boredom. (If you're decid-edly anti-spider, you might check out the 'Bonnie' cultivar, which has attrac-tive curly leaves.) A spider plant's leaves are generally green- and white-striped and arch outward from the center of the plant like a fountain. The plant sends out long stems, called *runners*, at the bottom of

which are spidery-looking babies, called *plantlets*. In nature, these plantlets would land on the soil and take root as new plants. Since there's probably no soil on your floor or table, you can pot these babies up—or just let them hang out.

How to Make Friends with Spider Plants

>> SOIL

Use an all-purpose potting mix.

>> LIGHT

Bright indirect light is best, such as that from an eastern or western window. Keep out of full sun to avoid sunburn; filter the light from southern exposure.

>> WATER AND HUMIDITY

Water when the top half inch (1 cm) of the soil feels dry. Occasional misting will make your plant happy, especially in dry rooms. The spider plant can be a little sensitive to water that contains fluoride, which can cause its leaf tips to turn brown. If your water comes from a municipal source, consider using filtered water.

>> TEMPERATURE

Room temperatures of 55–80°F (13–27°C) are suitable for this plant.

>> SIZE

A spider plant is usually around 12 inches (30 cm) tall.

>> BUGS AND DISEASE

Watch out for root rot (page 44), spider mites (page 42), scale (page 41), and whiteflies (page 42).

>> OTHER TIPS

To keep your plant looking natty, remove brown leaf tips with sharp, clean scissors.

If you want the plant to produce babies, don't overdo it with the fertilizer. Fertilizer can also turn the leaf tips brown, so occasionally water your plant in a sink to flush out excess salts.

STAGHORN FERN (PLATYCERIUM BIFURCATUM)

If you're looking for unusual plants to fill your collection, look no further. With silvery-green felted fronds that resemble antlers, the staghorn fern (also called the *elkhorn fern*) is no dainty, lacy fern. You'll often find staghorns mounted on wooden boards on a bed of sphagnum moss, like some kind of plant trophy, but they can also be grown in mesh or wire baskets—a much easier option in terms of watering and display. The base of the plant is composed of round leaves called shield fronds, which cover the roots that attach to whatever surface the fern is growing on. The shield fronds turn brown over time, so don't make the mistake of thinking that they're dead: just leave them alone. Like many orchids and bromeliads, the staghorn is an epiphyte (an air plant). It's relatively easy to care for, as long as you can satisfy its need for humidity.

How to Make Friends with Staghorn Ferns

» SOIL

Staghorn ferns often come mounted on boards, usually on a bed of sphagnum moss. If you'd prefer to put yours in a hanging basket, line the basket with a mix of sphagnum moss and an orchid potting mix, and then add the fern.

≫ LIGHT

The key to success with a staghorn fern is moderate to bright indirect light, such as that from an eastern window. Never put it in full sun or the leaves will get sunburned.

≫ WATER AND HUMIDITY

Staghorns are fairly tolerant when it comes to temperature, but they do want humidity. Mist your plant, especially if the room is dry, or consider placing it near a humidifier or in a higher-humidity room like a bathroom. Don't let the potting medium dry out completely before watering the root ball at the base of the plant. You can also soak the roots briefly in water. You may need to water more often in hot, dry weather.

≫ TEMPERATURE

Room temperatures of 60–80°F (16–27°C) are suitable for this plant.

≫ SIZE

An staghorn fern's fronds can grow up to 3 feet (90 cm) long.

≫ BUGS AND DISEASE

Watch out for scale (page 41).

≫ OTHER TIPS

As with other ferns, avoid using chemicals on the sensitive fronds. If scale appears, remove it with a damp paper towel.

TREE PHILODENDRON (PHILODENDRON BIPINNATIFIDUM, SYN. PHILODENDRON SELLOUM)

With massive leaves that somewhat resemble a many-fingered open palm, the sprawling tree philodendron is a dramatic plant. If you're a fan of monstera (page 118), you might like this one as well, and it has similar needs. It will definitely give your home a tropical vibe, but you'll want to give it a lot of room because this "plantspreader" grows out as well as up. It can often end up wider than it is tall (like up to 6 feet/1.8 m wide), and the leaves can reach 3 feet (90 cm) long. Over time, the plant will develop a trunk and aerial roots. If you like the look of this plant, but not the size, check out a smaller cultivar like 'Xanadu'.

How to Make Friends with Tree Philodendrons

» **SOIL**

Use an all-purpose potting mix.

» **LIGHT**

The tree philodendron is quite flexible when it comes to light requirements. It will do very well in moderate to bright light (an east- or west-facing window), but will also adapt to lower-light locations, such as a northern window.

» **WATER AND HUMIDITY**

Water when the top inch (2.5 cm) of the soil feels dry. As a tropical plant, it likes warmth and humidity. If your home is dry in winter, a pebble tray (page 18) can be helpful.

» **TEMPERATURE**

Room temperatures of 60–80°F (16–27°C) are suitable for this plant.

» **SIZE**

Over time, this floor plant can get quite big. A large plant may be up to 6 feet (1.8 m) wide and several feet tall.

» **BUGS AND DISEASE**

Watch out for mealybugs (page 40), scale (page 41), and root rot (page 44).

» **OTHER TIPS**

Those big leaves attract dust. You can wipe them down with a damp cloth, or even give the plant a shower once in a while.

URN PLANT *(AECHMEA FASCIATA)*

Bromeliads are a group of epiphytic and terrestrial plants that includes air plants, pineapple plants, and many others. Bromeliads such as the urn plant are unusual in that they have sturdy leaves that form a central reservoir called a *vase* or *tank*. When watering, you'll add water to both the soil and the tank, as if you're filling up a pitcher. In its native habitat, bromeliads grow on or under trees, so this opening collects any rainwater that falls its way.

When it is three to four years old, your plant will produce a stunning pink and purple bloom that lasts several months. This happens only once, and then the plant slowly dies. Keep this in mind when you buy your plant, as they are often sold in bloom. But take heart! All hope is not lost. This natural process will produce several offsets, or pups, that can be repotted when they are about 6 inches (15 cm) tall.

How to Make Friends with Urn Plants

≫ **SOIL**

Use a bromeliad mix if you can find it; otherwise, use an orchid mix.

≫ **LIGHT**

Urn plants like bright and/or filtered light from an east- or west-facing window.

≫ **WATER AND HUMIDITY**

Add water both to the tank and to the potting mix. Keep an inch or two (2.5–5 cm) of water in the tank. You'll need to change this water every few weeks. You can also give the plant a little shower to rinse any dust off its leaves. This plant is a little sensitive to water quality, so use filtered or distilled water. As a tropical plant, the urn plant likes humid conditions; use a pebble tray (page 18) to boost humidity.

≫ **TEMPERATURE**

Room temperatures of 60–80°F (16–27°C) are suitable for this plant.

≫ **SIZE**

An urn plant is usually 1–2 feet (30–60 cm) tall.

≫ **BUGS AND DISEASE**

Watch out for root rot (page 44) and scale (page 41).

≫ **OTHER TIPS**

After the bloom dies, cut it off and continue taking care of the parent plant (R.I.P.) until the pups are ready to be repotted.

WATERMELON PEPEROMIA (PEPEROMIA ARGYREIA)

Undemanding and low key, peperomias are popular, easy-to-find foliage plants. Take one look at a watermelon peperomia and you'll understand how it got its name. The pattern on the tear-shaped leaves evokes a watermelon rind and—quite appropriately—the stems are red. If you like the Chinese money plant (page 82), you might also like the watermelon peperomia, as it has a somewhat similar cute look. Originally from South America, these plants stay fairly compact, making them a great choice for smaller spaces. While a single small plant might look a bit lonely on its own, you can pair it with a few other petite companions on a windowsill.

How to Make Friends
with Watermelon Peperomias

» **SOIL**

Use an all-purpose potting mix. You can add a little perlite for
extra drainage to keep the roots from staying too wet.

» **LIGHT**

Watermelon peperomia likes bright filtered or indirect light, such
as that from an eastern window. Keep it out of direct sunlight.

» **WATER AND HUMIDITY**

Water when the top half inch (1 cm) of the soil feels dry. This
tropical plant doesn't like to dry out too much, but it also doesn't
like to be overwatered, as root rot can set in. Water less frequently
in fall and winter. If your home is very dry, you can boost humidity
by using a pebble tray (page 18), particularly in summer.

» **TEMPERATURE**

Temperatures of 60–80°F (16–27°C) are suitable for this plant.

» **SIZE**

This small plant is usually around 6–8 inches (15–20 cm) tall.

» **BUGS AND DISEASE**

Watch out for signs of root rot (page 44), mealybugs (page 40), or
spider mites (page 42).

» **OTHER TIPS**

If you'd like to try your hand at propagation, peperomias are a
good choice. All you need to do is pot up a leaf cutting (page 50) in
a well-draining potting mix.

WAX PLANT (HOYA CARNOSA)

The wax plant (sometimes called the *wax flower plant*) has long been a favorite in the home, perhaps due to its indestructibility. Like most succulents, it can survive some neglect, so it's good for the forgetful houseplant owner or frequent traveler. If you want your wax plant to thrive and flower, make sure you give it bright light and don't fuss over it

too much. This plant even prefers to be slightly pot-bound, especially when producing its unique, scented flowers.

As a climbing or trailing plant, you can train it on a trellis or simply let its vines hang down from a tall pot or hanging basket; it's not fast-growing, so don't worry about it taking over your living room. If you find the standard *Hoya carnosa* a little too ho-hum, check out the 'Tricolor' cultivar, which has variegated leaves with white and pink edges.

How to Make Friends with Wax Plants

>> **SOIL**

Use a well-draining, all-purpose potting mix. You can add extra perlite to improve drainage.

>> **LIGHT**

A wax plant likes bright light with some direct sun, such as that from an east-facing window. In sunnier locations, provide some protection, like a sheer curtain, to avoid sunburn. You can grow this plant under lower light conditions, but it may not produce flowers.

>> **WATER AND HUMIDITY**

Water when the top inch (2.5 cm) of the soil feels dry. Water less frequently in winter.

>> **TEMPERATURE**

Room temperatures of 60–80°F (16–27°C) are suitable for this plant.

>> **SIZE**

A wax plant is a climbing/trailing plant, with vines that can grow several feet long.

>> **BUGS AND DISEASE**

Watch out for mealybugs (page 40) and root rot (page 44).

>> **OTHER TIPS**

Don't remove the flower stalk after the flowers fade; future flowers will bloom from that same spot.

You can easily propagate this plant with a stem-tip cutting (page 51).

ZEBRA PLANT (HAWORTHIA FASCIATA)

There is something irresistible about the diminutive zebra plant, often called simply *haworthia*. With its zebra-like white stripes, this wee succulent looks good with almost any decor and in any kind of pot; try a glossy gray ceramic pot, an industrial-looking concrete pot, or a petite terra-cotta container. Group a few zebra plants together in a bigger pot, or pair them up with a few other succulents to make your own curated garden. The tips are quite spiky, so watch your fingers!

A zebra plant is an excellent choice for a narrow windowsill or small spaces, and a godsend for anyone who forgets to water their plant, as it needs very little water.

How to Make Friends with Zebra Plants

>> SOIL

Use a cactus and succulent potting mix.

>> LIGHT

A zebra plant likes bright indirect light, such as that from an eastern or western window. If you want to put it in full sun, do so gradually to avoid sunburned leaves.

>> WATER AND HUMIDITY

Water when the top inch (2.5 cm) or so of the soil feels mostly dry. Water sparingly in winter.

>> TEMPERATURE

Room temperatures of 60–80°F (16–27°C) are suitable for this plant. It can tolerate cooler temperatures, down to 50–55°F (10–13°C), in the winter.

>> SIZE

This small plant grows to 3–4 inches (7.5–10 cm) tall.

>> BUGS AND DISEASE

Watch out for root rot (page 44) and sunburn (page 47).

>> OTHER TIPS

If your zebra plant doesn't seem to be doing much, don't worry: it tends to grow slowly.

Like the aloe plant (page 64), the zebra plant also produces babies (called *pups* or *offsets*), which can be removed and potted up individually.

ZZ PLANT (ZAMIOCULCAS ZAMIIFOLIA)

The ZZ plant's name comes not from any rock-band reference, but from its botanical name: *Zamioculcas zamiifolia*. Originally from eastern Africa, the ZZ—also called the *Zanzibar gem*—is extremely drought resistant and can tolerate low light levels and dry air. Like the snake plant (page 148), the ZZ is almost bulletproof. The less you do to it, the better. Its leaves continue to stay glossy green, no matter what you forget to do. My ZZ, for example, often goes weeks without a watering and doesn't complain. In addition to being easy to care for, a ZZ plant is quite beautiful, and its manageable dimensions make it perfect as a medium-sized floor plant next to a couch. Smaller ZZ plants can add interest to a table or pedestal.

How to Make Friends with ZZ Plants

>> **SOIL**

Use a cactus and succulent mix or a well-draining, all-purpose potting mix.

>> **LIGHT**

A ZZ plant can tolerate relatively low levels of light, such as that provided by a north-facing window, but it will thrive in bright indirect light, such as that from an eastern or western exposure. Avoid direct sun, which can cause sunburn.

>> **WATER AND HUMIDITY**

Wait until the top 2 inches (5 cm) of the soil feel dry before watering. If in doubt, wait a little longer, because the ZZ plant can handle drought. If it gets *too* dry, it can start to drop leaves; if it receives too much water, the leaves may start to turn yellow. It doesn't require a lot of humidity.

>> **TEMPERATURE**

Room temperatures of 60–80°F (16–27°C) are suitable for this plant.

>> **SIZE**

A ZZ plant grows to about 2 feet (60 cm) tall.

>> **BUGS AND DISEASE**

If you overwater your plant, it could become susceptible to root rot (page 44).

While it's always a good idea to wear gardening gloves while working with plants, it's especially true when working with a ZZ. It releases a sap that can irritate the skin.

Every once in a great while, haul your ZZ into the bathroom to give it a shower, to get rid of the dust that accumulates on its many leaflets. This is a quick-and-easy way to keep your plant clean, but you can also wipe the leaflets by hand with a damp cloth if the plant has become too heavy and unwieldy to carry.

A large ZZ plant can be divided and the divisions potted up separately.

ACKNOWLEDGMENTS

S pecial thanks to Barbara Berger for believing I'd be a good fit for this project and to Jennifer Williams for thinking of me. My editor, Elysia Liang, was the best reader one could hope for, and her comments, queries, and careful edits helped greatly to improve the manuscript and make it even more accessible. Thanks also to the whole team at Sterling, especially to production editor Scott Amerman, for keeping track of all the art and his careful attention to detail; David Ter-Avanesyan, who captured the spirit of the book in his lovely cover design; the interior designer, Shannon Plunkett, who carried that spirit throughout its pages; and Stephanie Augello and Cynthia Carris, for the big job of finding all of the right images. Extra-special thanks to my mother, Rowena Rodino, who knows much more about plants than I do, for planting a seed that may have taken a long time to germinate. And to my husband, Pierre Baumann, who supports me in all I do, including buying more plants.

RESOURCES

General Plant Information

MISSOURI BOTANICAL GARDEN (MissouriBotanicalGarden.org)
An excellent resource. Click on the Gardens & Gardening tab and use the
Plant Finder to get accurate, no-nonsense growing information on more than
7,500 plants, including houseplants, outdoor plants, and edible garden plants.
The plant profile pages even include a place to listen to the pronunciation of
the plant's botanical name.

THE PLANT LIST (ThePlantList.org) A collaboration of various botanical
gardens around the world, this website bills itself as a "working list of all
known plant species."

History

For a history of houseplants, check out Tovah Martin's *Once Upon a Windowsill:
A History of Indoor Plants* (Timber Press, 1988) and Catherine Horwood's *Potted
History: The Story of Plants in the Home* (Frances Lincoln, 2007).

Online Nurseries

LOGEE'S (Logees.com) Logee's stocks and ships an extensive collection of
rare, fruiting, and tropical plants to most parts of the U.S.

PISTILS (Shop.PistilsNursery.com) Based in Portland, Oregon, Pistils stocks and ships many popular plants. They also have pots and other gardening accessories and offer detailed advice on plant care on their website.

THE SILL (TheSill.com) Based in NYC, The Sill ships plants, pots, and accessories nationwide.

Plant Problems

For a thorough explanation and helpful photos of almost anything that can go wrong with your houseplant, read David Deardorff and Kathryn Wadsworth's *What's Wrong with My Houseplant?* (Timber Press, 2016).

Podcast

Jane Perrone, formerly of *The Guardian*, hosts a terrific podcast about houseplants called *On the Ledge*. Listen through her website, JanePerrone.com, or download episodes via your favorite podcast app.

Poisonous Plants

ASPCA TOXIC AND NON-TOXIC PLANTS LIST (https://www.aspca .org/pet-care/animal-poison-control/toxic-and-non-toxic-plants) Visit this website for a database of plants that may be toxic to your dog or cat, as well as those that are safe to keep around your furry friends. You can search by either common or scientific name.

NATIONAL POISON CONTROL CENTER (https://www.poison.org /articles/plant) The National Poison Control Center maintains a database of poisonous plants, including photographs.

Pots and Garden Tools

For unique pots and garden tools, check out Etsy (Etsy.com), Terrain (ShopTerrain.com), IKEA (Ikea.com), and West Elm (WestElm.com). Many Etsy shops also sell live plants.

BIBLIOGRAPHY

"Air Plant Care." PistilsNursery.com. January 19, 2015. https://shop
.pistilsnursery.com/blogs/the-care-blog/18673779-air-plant-care-how-to
-care-for-air-plants-aeriums-and-tillandsia-mounts.

"'Bullet-Proof' Houseplants: Perfect for Low Light." GardensAlive
.com. Accessed June 25, 2018. https://www.gardensalive.com/product
/bulletproof-houseplants-perfect-for-low-light.

Cruso, Thalassa. *Making Things Grow Indoors*. New York: Knopf, 1969.

Darwin, Charles, to Asa Gray. Darwin Correspondence Project, "Letter no.
3662." Accessed June 25, 2018. https://www.darwinproject.ac.uk/letter
/DCP-LETT-3662.xml.

Deardorff, David, and Kathryn Wadsworth. *What's Wrong with My
Houseplant?* Portland, OR: Timber Press, 2016.

Dong, Qianni. "Problems Common to Many Indoor Plants."
Missouri Botanical Garden. Accessed June 25, 2017. http://www
.missouribotanicalgarden.org/gardens-gardening/your-garden/help-for-the
-home-gardener/advice-tips-resources/visual-guides/problems-common
-to-many-indoor-plants.aspx.

Hodgson, Larry. *Houseplants for Dummies*. Hoboken, NJ: Wiley, 2006.

Husted, Kristofor. "Can Gardening Help Troubled Minds Heal?" NPR.org, February 22, 2012. https://www.npr.org/sections/thesalt/2012/02/17/147050691/can-gardening-help-troubled-minds-heal.

Lee, Min-Sun, Juyoung Lee, Bum-Jin Park, and Yoshifumi Miyazak. "Interaction with Indoor Plants May Reduce Psychological and Physiological Stress by Suppressing Autonomic Nervous System Activity in Young Adults: A Randomized Crossover Study." *Journal of Physiological Anthropology* 34, no. 1 (2015): 21. doi: 10.1186/s40101-015-0060-8. https://www.ncbi.nlm.nih.gov/pmc/articles/PMC4419447.

Livni, Ephrat. "The Japanese Practice of 'Forest Bathing' Is Scientifically Proven to Improve Your Health." *Quartz*, October 12, 2016. https://qz.com/804022/health-benefits-japanese-forest-bathing.

Martin, Tovah. *The Indestructible Houseplant*. Portland, OR: Timber Press, 2015.

Mason, Sandra. "Soil Conditioners Are Explained." The Homeowners Column, University of Illinois Extension. Accessed June 29, 2018. https://web.extension.illinois.edu/cfiv/homeowners/981128.html.

"Novice Paphiopedilum Culture Sheet." American Orchid Society. Accessed June 25, 2018. http://www.aos.org/orchids/culture-sheets/novice-paphiopedilum.aspx.

Park, Seong-Hyun, and Richard H. Mattson. "Ornamental Indoor Plants in Hospital Rooms Enhanced Health Outcomes of Patients Recovering from Surgery." *Journal of Alternative and Complementary Medicine* 15, no. 9 (September 2009). doi: 10.1089/acm.2009.0075. https://www.ncbi.nlm.nih.gov/pubmed/19715461.

Perrone, Jane. "Orchids: Sow, Grow, Repeat Winter." Sow, Grow, Repeat. Podcast produced by Rowan Slaney. *The Guardian*, February 6, 2016. https://www.theguardian.com/lifeandstyle/audio/2016/feb/06/orchids-sow-grow-repeat-winter.

"Plants Clean Air and Water for Indoor Environments." NASA Spinoff. Accessed June 29, 2018. https://spinoff.nasa.gov/Spinoff2007/ps_3.html.

Pleasant, Barbara. *The Complete Houseplant Survival Manual.* North Adams, MA: Storey, 2005.

Ramsey, Ken. "Sphagnum Moss vs. Peat Moss." National Gardening Association, July 22, 2014. https://garden.org/ideas/view/drdawg/1972/Sphagnum-Moss-vs-Peat-Moss.

"Rooting Cuttings in Water." Missouri Botanical Garden. Accessed August 22, 2018. http://www.missouribotanicalgarden.org/gardens-gardening/your-garden/help-for-the-home-gardener/advice-tips-resources/visual-guides/rooting-cuttings-in-water.aspx. .

"Sex and Lies," *Plants Behaving Badly*. Directed by Steve Nicholls, narrated by David Attenborough. PBS, 2013.

Stuart-Smith, Sue. "Horticultural Therapy: 'Gardening Makes Us Feel Renewed Inside." *The Telegraph*, May 31, 2014. https://www.telegraph.co.uk/gardening/10862087/Horticultural-therapy-Gardening-makes-us-feel-renewed-inside.html.

Vincent, Alice. "Succulents to Share: How to Propagate Houseplants in Glass." *The Telegraph*, February 9, 2018. https://www.telegraph.co.uk/gardening/how-to-grow/succulents-share-propagate-houseplants-glass/.

Wolverton, B. C., Willard L. Douglas, and Keith Bounds. "A Study of Interior Landscape Plants for Indoor Air Pollution Abatement." NASA, July 1989. https://archive.org/details/nasa_techdoc_19930072988.

 IMAGE CREDITS

Alamy: De Agostini / G. Cigolini 80; Garden World Images Ltd 78; Steffen Hauser/botanikfoto 98; Fir Mamat 136; RF Company 112; DebraLee Wiseberg 50

Dreamstime: 88and84 back cover, 84; Artjazz 132; Bozhenamelnyk 88; Sakesan Khamsuwan 19; Sharaf Maksumov 4; Pinonsky 103; Shoeke27 62; Ivonne Wierink 148; Linda Williams 126; Zhaojiankang 121

Flickr: Scot Nelson 41, 43, 45

Getty: DEA / C. DANI 154; ML Harris 114; Sian Irvine 134

iStock: 200mm spine, 64; Автор 101; KatarzynaBialasiewicz 7, 53, 118; brizmaker 37; Cleardesign1 30; Dar1930 60; FatCamera 25; fontgraf front cover; gyro 34; FeelPic 140; franhermenegildo 146; insjoy 92; Artem Khyzhynskiy 142; Bogdan Kurylo 36; LightFieldStudios front cover; loonara ii, x, xiii-xiv, 1, 58-59; OlgaMiltsova spine, 71, 124; Andrey Mitrofanov 46; Mykeyruna front cover; Andrey Nikitin 108; Noppharat05081977 40; timnewman 152; Chansom Pantip 48; OlgaPonomarenko 23; Povareshka vi-ix; kevinruss front cover; sagarmanis 3; AnikaSalsera back cover, 94; SawBear 162; Socha 76; Voisine 86

Shutterstock: 826A IA 46; Abra Cadabraaa endpapers; Tatyana Abramovich 74; Amilao 130; Aunyaluck 156; Mary Berkasova 69; Catherine 311 42; AJCespedes 45; Del Boy 110; Demkat 28, 90; Isabel Eve 47; Floki 42; flowerstock 2; Shelsea Forward 96; fottodk 19; Val_Iva 20; JRP Studio 33; rattiya lamrod 28; LAVRENTEVA 54-57, 174; Luoxi 51; Alexander Maksimov 160; mart i, xi, 9, 31-32, 35, 60-166, 168, 178; mates 21; Bozhena Melnyk 150; Patrycja Nowak back cover, 82; Sarycheva Olesia 66; Photo and Vector 164; PHOTO FUN 39; Saturn29 128; Maria Sem v, 24, 38, 49, 167, 170; Stanislav71 44; Anna Sulencka2 144; SzB 138; tdee photo cm 158; thefoodphotographer 106; VikaValter 12; Merkushev Vasiliy 16

INDEX

Page numbers in italics indicate plant profiles.

ABOUT THE AUTHOR

HEATHER RODINO grew up in rural Pennsylvania surrounded by plants and gardens but has lived in urban locations for most of her adult life, making small-space and indoor gardening a necessity. A writer and editor, she is the author of several books and has worked for many years in the publishing industry. She lives in San Juan, Puerto Rico.